100 Woods

Please Return to
Robert Easton

100 Woods

A Guide to Popular Timbers of the World

Peter Bishop

The Crowood Press

First published in 1999 by
The Crowood Press Ltd
Ramsbury, Marlborough
Wiltshire SN8 2HR

British Library Cataloguing-in-Publication Data
A catalogue record for this book is available from the British
Library.

ISBN 1 86126 167 5

Artwork credits:
Figures 1, 3, 4, 5 & 6;
Department of the Environment,
Forest Products Research,
Bulletin No. 56
(Revised edition of FPR Record No.21),
The Growth and Structure of Wood
by B. J. Rendle, revised by J. D. Brazier,
HMSO, 1971 (ISBN 11 470543 7)

Typeset in: New Baskerville and Optima

Typeset and designed by Focus Publishing, Sevenoaks, Kent

Printed and bound by China

CONTENTS

FOREWORD

There is increasing public awareness of the appeal of wood as a raw material for building and furniture manufacture and an equally increasing awareness of threats to many timber species through over-exploitation and inefficient use. This book seeks to explain in simple terms the nature of wood and its principal components and then to tabulate for the most important species worldwide the significant group of properties. A list of common and Latin scientific names is also provided. The book compiles in one place and in easily interpretable form a large amount of information about an impressive number of species; it should be of interest to all those concerned with the use of wood or appreciative of wood as a raw material.

Professor Jeffery Burley
Director
Oxford Forestry Institute
July 1998

The Oxford Forestry Institute, OFI, is internationally recognized as one of the leading centres of excellence in forestry and related subjects; it is based at the Department of Plant Sciences, University of Oxford. Professor Jeffery Burley has been its Director for a number of years. He travels widely throughout the world lecturing on all subjects related to wood; he is recognized as one of the major authorities in this field.

INTRODUCTION

I have often asked myself what it is that I find so fascinating about wood. I like trees and appreciate their form and beauty, but it is the raw material they produce that, in my opinion, is so interesting. Each piece is unique. Although I have tried, often inadequately, to describe the 'typical' features of the woods in this book, words cannot compare with the actual feel, appearance and smell of individual timbers. People always seem to want to touch wood, which testifies to the strength of its allure. The trouble is that there is less and less of the real stuff around these days, what with all the composite boards and pressed paper veneers now widely available. However, we are utilizing timber as a raw material so much better than we used to. There is now less waste than ever before and more appreciation that, although wood is a renewable resource, we need to take care of it.

Let's return to that question, why wood? I was actually more interested in metal work and technical drawing than wood when I was at school. I was also lazy. In those days if you did not show any signs of academic interest the career's officer told you straight: 'Get a job, lad'. So I did. At fifteen I left the flat, bustling plains of north-east Oxford and ended up in the rural tranquillity of West Herefordshire. Nestling in this beautiful countryside was a timber yard owned and operated by a distant uncle. In those early days it was a hard life, and I was poorly paid and treated. In fact I have no idea why I stayed – it must simply have been an in-built pig headedness. Any signs of laziness quickly disappeared: I had to work hard – there was no choice. Living on the breadline, I moonlighted making joinery and furniture, the only way I knew how to supplement my income. It was in these early days that I first appreciated how forgiving and flexible wood is as a material. There are not that many mistakes made that cannot be corrected. This harsh introduction formed the basis of the love affair that continues today!

Time moved on. I have enjoyed inspecting parcels of wood from the clubs of Liverpool, through Europe and on to exotic locations in the Far East. Along the way I have met some interesting and wonderful people, many who deserve a book of their own. I have made friends, been helped and encouraged by the great and the good, and come across a few so-and-so's as well! Like all industries the timber and associated trades are incestuous. There is not much going on that someone does not know about, or knows a man who does. Perhaps this is another reason why I find wood interesting.

I recently set off on a voyage of rediscovery, and this book is the result. At some time I have handled and worked with most of the woods described here, although some not for many years. Researching each was fun. Where appropriate I have included a personal comment that may help in the appreciation of a particular wood. I was privileged to spend part of a day at the Oxford Forestry Institute, University of Oxford, with Ian Gourlay. He allowed me to burrow in the bowels of the institute, wander the corridors and view the lecture rooms where one is constantly surrounded with examples of beautiful wood from around the world. This is an experience not to be forgotten and I hope to revisit sometime in the near future. Ian is the curator at the OFI and is responsible for the excellent photographs contained herein. The OFI have produced a CD-ROM database of wood properties: look under 'Further Information' on page 218 for details.

Having got your hands on a copy of this book, I hope you will find it interesting and of value.

ABOUT WOOD

Growth and Structure

Trees are large, woody plants having a cellular structure that grows through division of specific cells contained within the outer cambium layer below the bark (see Fig. 1). Trees differ from other woody plants by having one main trunk or stem. This stem provides the strength of the whole, transporting minerals contained in water gathered by the roots to the branches and leaves. Each of the main sections of the tree, crown, trunk and roots has a specific task to perform.

The crown, using sunlight and through the process known as photosynthesis, absorbs carbon dioxide from the air. This is mixed with water containing minerals from the roots to produce carbohydrates which then form the cellular structure. Apart from the food gathering function of the roots they also act as a safe anchor from which the tree can develop and grow. The trunk acts as the conduit between these other two main components, providing strength and food storage facilities as well (see

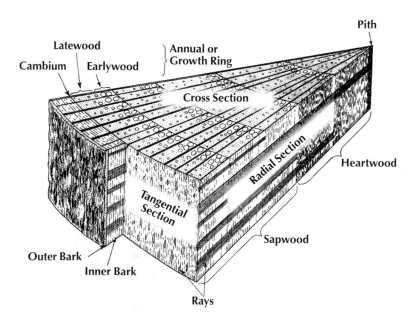

Fig. 1 *Diagrammatic drawing of a wedge-shaped segment cut from a five-year old stem of a hardwood, showing the principal structural features.*

Fig. 2). The beneficial effect for humanity from this process is probably threefold: removing carbon dioxide from the atmosphere and releasing oxygen; creating a necessary haven for other plants, mammals and insects; and, finally, when managed correctly providing a truly sustainable source of raw material.

Growth through division of the cambium cells can be continuous or seasonal. In temperate climates the spring or 'early' growth tends to be rapid and apparent. This slows throughout the year until it halts during the winter period; the cycle is then repeated in following years. This seasonal growth creates the typical patterns or rings that are visible to the naked eye. Under magnification the ring system becomes clearer (see Figs. 3 and 4).

Fig. 2 *Growth of a tree.*

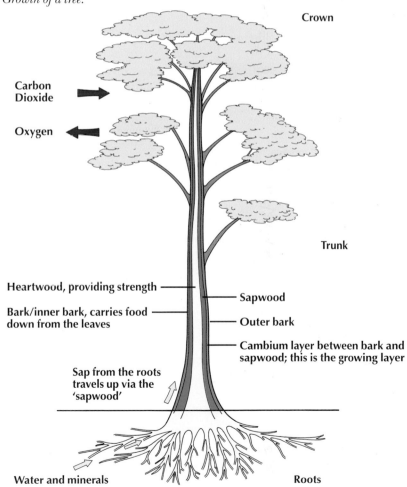

Crown

Carbon Dioxide

Oxygen

Trunk

Heartwood, providing strength

Bark/inner bark, carries food down from the leaves

Sapwood

Outer bark

Cambium layer between bark and sapwood; this is the growing layer

Sap from the roots travels up via the 'sapwood'

Water and minerals

Roots

Hardwood trees that grow in this way are known as having a 'ring porous' structure. In tropical climates there tends to be little or no seasonality to interrupt the cycle, with a few exceptions. Hardwoods from these regions produce a 'diffuse porous' structure (see Fig. 5). The growth of a tree can be likened to the act of continuously pulling a glove onto your hand, and then another, layer upon layer, until this function ceases for whatever reason.

The functions of the cellular structure can, in a similar way to the overall tree form, be divided into three main tasks: mechanical, transportation and storage of foodstuff. The main strength properties in softwoods come from the thickened walls of the latewood tracheids. Foodstuffs are transported through the specially designed early wood tracheids that are interconnected where they overlap and also through a series of 'pits' in these cell walls. Storage of foodstuffs is carried out by the final element, parenchyma cells, that are common to both softwoods and hardwoods (see Fig. 6). These are a series of cells that run both vertically and horizontally within the structure. The horizontal cells are sometimes visible to the naked eye in some hardwoods and help to create patterns or 'figure'. Most of the parenchyma cells will need some form of magnification to identify them. In hardwoods fibres form the main strength constituent. These are generally thick-walled cells distributed throughout the structure and make up the bulk of the woody tissue. Vessels are the cells that act as the conduit for the foodstuffs. These are usually visible to the naked eye or under slight magnification; conversely, softwood tracheids will

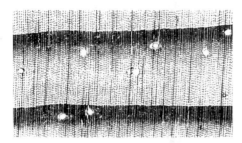

Fig. 3 *Scots Pine or Redwood.*

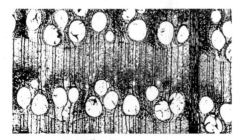

Fig. 4 *Oak.*

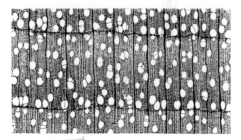

Fig. 5 *Birch.*

need microscopic magnification before they become clearly visible. Combinations of these visual cell structures associated with both hardwoods and softwoods will provide some of the 'gross' features needed as aids to identification.

Softwood or Hardwood?

Apart from their peculiar cellular structures discussed above, there are some fundamental differences between softwoods and hardwoods. These differences are not related to the weight, strength or hardness of each group – for example, Balsa is a hardwood and Pitch Pine a softwood! Both are extremes within their classifications.

Most commercial timbers will be divided into one of the two classifications. Softwoods are contained within the grouping known as gymnosperms. Broadly, these are

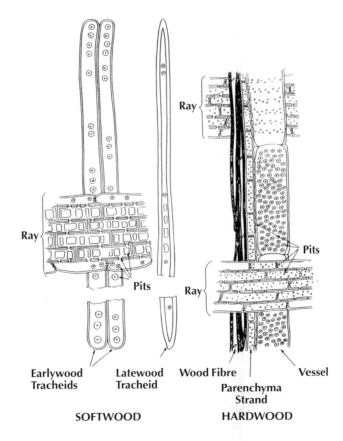

Fig. 6 *The principal features of a softwood and a hardwood as seen in radial section. Only one-third of the length of the softwood tracheids is shown. (Magnification x 100 approx.).*

trees that have needle-shaped leaves that are generally evergreen and are cone-bearing, with naked seeds. Conversely, hardwoods are angiosperms. These are trees that have broad leaves with encased seed pods. Most hardwoods in temperate regions are deciduous, shedding their leaves during autumn; in the tropics they are generally evergreen. Some softwoods, such as Larch, are also deciduous, but in the main these tend to be the exception rather than the rule.

Classification

Achieving botanical correctness when naming timbers is never easy, and no doubt more controversy will be generated by this book. For clarity, if that is possible, an outline of how each of the woods is classified might prove useful. Early on, some sort of order was needed to name plants, although it was recognized that no two plants are ever the same. Reproduction of groups with similar features, over successive generations, and with only slight variations, have been separated into species. Some of these species share other common features and this provided the further botanical concept of a genus. This has led to the standard two-part nomenclature of plants. Initially, the name identifies botanically to which genus the plant belongs and the final part refers to the species. In this book, for example, you will find that 'Beech'

(*Fagus sylvatica*) provides definitive genus and species information, while 'Keruing' (*Dipterocarpus spp.*) defines the genus but contains a mixed bag of species. Although there is often some dispute among botanists regarding definitive classification of some plants, all are identified by one botanical name given in two parts. This is the universally accepted standard practice and the descriptions are always in Latin.

As a rule, D. J. Mabberley's *The Plant Book*, available from the Cambridge University Press, has been used as the basis for classification of the individual timbers in this book. Information concerning the vernacular, trade and other common names have been gleaned from a wide variety of sources. These names are comprehensively detailed within the Index to aid the reader with some of the more obscure ones.

Working Properties – The Key

Associated with each timber in this book is a key providing information on working properties, durability and weight. The choice of headings has been limited to provide, in my opinion, the most pertinent. Because of wood's very nature, where it is grown, how it grows, for how long, and so on, this is not a precise science, but more an art! A piece of lumber cut from one particular tree may display a wide variety of different grain patterns, colours and weight

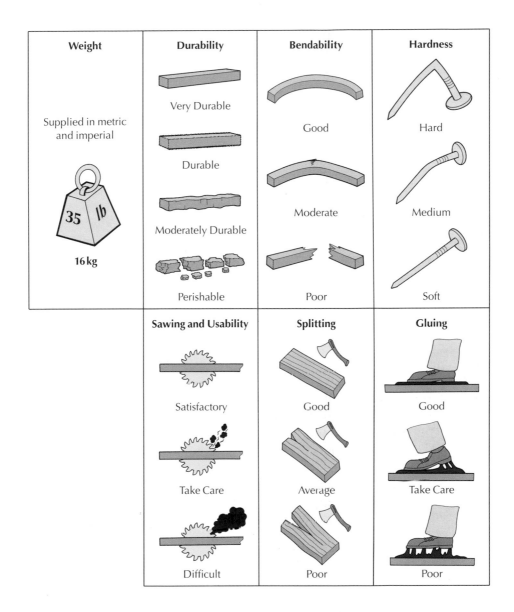

Weight	Durability	Bendability	Hardness
Supplied in metric and imperial	Very Durable	Good	Hard
	Durable	Moderate	Medium
35 lb	Moderately Durable		
16 kg	Perishable	Poor	Soft

	Sawing and Usability	Splitting	Gluing
	Satisfactory	Good	Good
	Take Care	Average	Take Care
	Difficult	Poor	Poor

variations. The key, therefore, should only be used as a guide, providing, as best it can, an indication of what each timber is like.

The full range of classifications are shown in the key above.

Weight

This is based on an average dry weight of between twelve to fifteen per cent moisture content. Moisture in wood is contained in two main areas, within the cellular cavities and in the cell walls.

The first is called 'free' moisture and is relatively easy to remove; the second is known as 'bound' moisture and needs some encouragement to vacate. Growth rates affecting the density of each tree will influence overall weights.

Durability

Timber is most susceptible to fungal attack and degradation at any point where a fair amount of moisture and air of the right temperature meet. Over time, individual timbers in various countries have been subjected to the 'graveyard' test, in which a sample of heartwood is placed into the ground and monitored for a number of years to see how long it resists attack by fungi. Where the wood comes from, its density, growth rates, and so on, will have some bearing upon the results. In some joinery, the initial moisture content of the wood being worked will become an influencing factor. If an unacceptably high level is present and is then locked in by paint, it is likely that the manufactured article will not last all that long, although fortunately today's paints and finishes tend to allow the timber to breathe. When deciding upon which timber to use for a specific job, some thought should be given to its suitability as well as its aesthetic values.

Bendability

All wood is elastic, a property that allows it to return to its original shape after being placed under stress; when the wood sheers, elasticity has failed. Normally, this classification refers to how much deflection is possible before the sample sheers, although in this book I have tempered this with reflections that provide a less scientific basis, in particular referring to how the timber will perform when steam-bent. The type of grain configuration, number of knots, and so on, will all be influencing factors.

Hardness

Hardness is generally linked to weight, although this is not always the case. The test used for this book has been to see how easy it is to penetrate the wood with a sharp object. Using a softwood for flooring is ill-advised. Conversely, using a hardwood for carcasing in furniture is unnecessary unless in a position of wear.

Sawing and Usability

An important factor is how easy the timber is to work with tools, whether by hand or with machines. Grain configurations will affect how likely it is that a saw will bind or how easily a surface will plane. If the timber has been abused or dried incorrectly this will also have some impact. This heading aims to provide a general indication of what to look out for. A fair proportion of woods will have 'take care' associated with them. This is the literal meaning until you can personally verify how to handle the particular piece or parcel of timber.

Splitting

Most woods will split if forced to, some more easily than others. In general, hardwoods will tend to split more readily than soft, although this is not always the case. In this classification, I have aimed to indicate whether the timber needs pre-drilling before fixing. I would recommend this with the majority of woods, especially if the fixing is near the end grain.

Glueing

Glue bonds pieces of timber together by penetrating into the cell structure of the adjoining faces, thereby forming a key into each. Many glues are stronger than the wood to which they are applied and most woods will bond well. It is only those with some inclusive factor, such as an oil, that do not.

Information for these classifications has been gathered from a wide variety of sources. However, the tables do not aim to be comprehensive, and if more technical information is required a visit to your local reference library should prove fruitful. Having personally handled and worked with the majority of the timbers in this book, I have tried to relate that knowledge, where applicable, within each description.

Source Countries

The world map key gives an indication of the source of each timber. Many species will be indigenous to a particular country or continent; others will have been introduced as plantation stock. I have used my knowledge and other sources of reference to indicate where each is sourced. This is not meant to be definitive, but provides information on where larger commercial quantities may be found.

Health and Safety Note

Dust from, and chemicals contained within, some timbers can cause irritation to the eyes, nose or throat. Additionally, there is a possibility that the occurrence of dermatitis and other skin infections is exacerbated by association with timber. These conditions are not general, but every precaution should be exercized to ensure that your health is protected. Take preventative action before embarking upon working with wood to avoid any possibility of risk. No specific comments have been associated with the timbers discussed in this book. The subject is dealt with comprehensively by the appropriate information available from, in the UK, the Health and Safety Executive. A summary and extracts are contained in the appendix, with details of how to acquire full information if required.

ABURA
Mitragyna spp.

Also known as: Bahia, Elelom, Elilom, Maza, Mujiwa, Mushiwa, Subaha, Voukou or Vuku.

Commercial supplies come mainly from the rainforests of West Africa, although Abura can also be found outside these zones. The timber is pale pinkish brown to light yellowish brown in colour. There is little or no obvious difference between the heart and sapwood; occasionally there is a small greyish core of heartwood. It is straight grained with a tendency to have some spiral or interlocking present. Although Abura is a plain wood in appearance, the occurrence of a wavy grain can at times produce a Birch-like figure. When green, this timber can produce a strong, slightly unpleasant smell that disappears after drying.

Abura is a fairly light wood, averaging 16kg (35lb) per cubic foot when dry. It dries well with a slight tendency to split and, as one would expect, shrinkage from green to dry can be more noticeable. It is considered a stable timber when dried, with little subsequent distortion or movement. Abura is rated as slightly weaker in strength than Beech, but is just a bit stiffer with more resistance to compression along the grain. The heartwood is classified as perishable and is moderately resistant to preservative treatment. The sapwood is reasonably permeable and it can be pressure-impregnated with preservative relatively easily.

Abura works reasonably well both by hand and machine, except for some abrasion and the occurrence of the spiral or interlocked grain. If any quantity is to be machined, then tungsten carbide tipped tools will be an advantage to avoid blunting. All tools should be kept sharp enough to

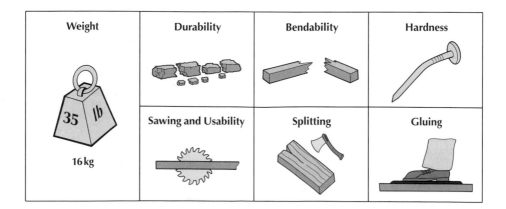

Weight	Durability	Bendability	Hardness
35 lb 16 kg	Sawing and Usability	Splitting	Gluing

ensure that there is little tearing out of the spiral grain, otherwise a woolly surface finish will result. Abura nails and screws satisfactorily but can have a habit of splitting, especially near the end grain; pre-drilling is therefore advisable. It is a good gluer, and can take stains well. A reasonable polished finish can be achieved.

With an in-built resistance to acid, Abura has historically been used in battery boxes and barrels. A first-class general purpose hardwood, with little movement, it has consistently been used for furniture and mouldings, while its acceptance of stain makes it useful as an edge-matching trim for decorative veneer-faced furniture. However, its lack of character means it is not often used as a show wood.

Smoking the leaves of Abura is more dangerous than opium!

Radial

Tangential

Source:

Predominantly West Africa.

Cross Section

AFRORMOSIA

Pericopsis elata (Formerly: *Afrormosia elata*)

Also known as: Bonsamdua, Devil's Tree, Kokrodua or Red Bark.

This is another tropical timber from West Africa. The local names of Devil's Tree or Red Bark derive from the blotches of red and orange that occur on the trunk of the living tree. Afrormosia is a beautiful golden brown timber with some yellow and reddish streaking apparent when first cut, changing progressively until eventually it becomes deep brown. The grain characteristics are interesting – some straight but the majority are interlocked which can cause problems with finishing. When the lumber is wet Afrormosia has a distinct but not unpleasant musky smell. Contact with ferrous metals when wet or damp will produce dark stains similar to those found on Oak.

A strong, heavy timber at around 21kg (46lb), on average per cubic foot when dry, it seasons fairly slowly without too much degrade apart from some distortion due to the interlocked grain. There is minimal shrinkage during the drying process, although some end splitting can occur. As one would expect, the relative strength of Afrormosia is greater than Beech. It is classified as very durable and is resistant to preservative treatment.

Afrormosia is an abrasive material, therefore tungsten carbide tipped tools should be used for sawing and planing, especially if more than a few pieces are to be machined. The interlocked grain can tear out when planing. Spelshing and burning can also occur when cross-cutting, so keep your tools sharp! Pre-drilling when nailing and screwing is necessary, otherwise expect the timber to split, especially at the ends. Afrormosia

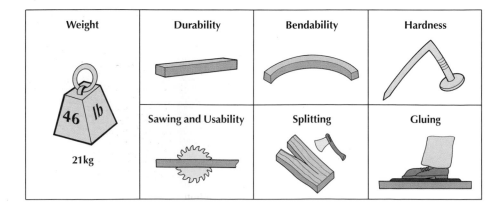

Weight	Durability	Bendability	Hardness
46 lb / 21kg			
	Sawing and Usability	Splitting	Gluing

takes glue well even though it appears to have a slightly oily surface feel. It also stains well, but probably better with spirit-based rather than water-based. A first-class finish can be obtained on this material producing a high gloss if desired.

This is an excellent timber that has often been called African Teak because of perceived similarities. It can easily be substituted for Teak when a cheaper material is required, and can be used for both internal and external joinery because of its good durability and stability. It is also popular within shop-fitting circles for bar and counter tops, doors, framing, mouldings and other similar applications. In furniture it is used as a show wood rather than for framing, and its durability and hard wearing properties also lend themselves to use as a flooring material.

Source:

Predominantly West Africa.

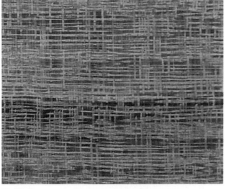

Radial

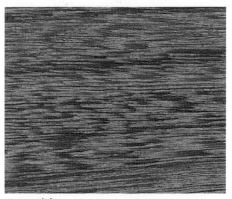

Tangential

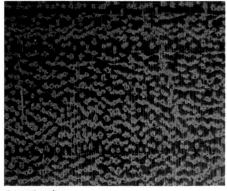

Cross Section

AFZELIA

Afzelia spp. (a mixture of *Afzelia* species: *A. africana,*
A. bipindensis, A. pachyloba and *A. quanzensis.*)

Also known as: Alinga, Apa, Aryian, Bilinga, Bolengu, Chanfuta, Doussie, Kontah, Lingue, Malacca Teak, Mbembakofi, Mkora, Mussacossa, Papao or Uvala.

Found throughout West, Central and East Africa, this timber has a fairly course grain with some interlocking evident. Coloured yellow to reddish brown initially, it turns dark red-brown on exposure. There is a clear definition between the sapwood, straw coloured, and the heartwood. Some mottled figure is apparent on occasions but the appearance is generally bland. Evidence of a white or yellow calcium-like deposit can be seen regularly within the pores. These deposits can leach out when the timber is wet or damp.

Because of the range of species weights can vary, but a general dry average would be around 21kg (46lb) per cubic foot. It dries slowly but reasonably well without too much distortion, although it does have a tendency to surface check and existing splits can be extended. Shrinkage is small and movement is minimal. This timber is considered exceptionally stable when dry and has been compared to Teak in that respect. Its strength values for the best average will equal and may well exceed those of Beech. Afzelia is very resistant to pressure preservative treatment, which is not surprising as it is classified as very durable – even termites do not like it!

Rated as moderately hard, this timber saws and planes reasonably well but may require tungsten carbide tipped tooling to maintain an edge. Pre-drill any nailing and screwing

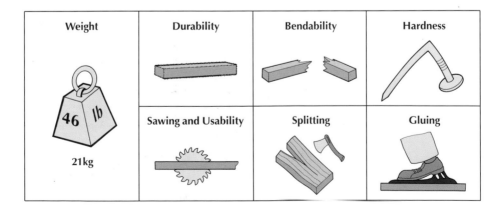

Weight	Durability	Bendability	Hardness
46 lb 21kg			
	Sawing and Usability	Splitting	Gluing

operations to avoid splitting. Gluing can be difficult as some species exude resin. If a stain is required, a spirit-based is best, especially on those components displaying yellow or white deposits. Surfaces can be difficult when the grain is interlocked, but with effort an excellent finish can be achieved.

Because of its high level of durability and reasonable strength properties this timber is popular for use in heavy construction work, especially in dock and quayside applications. Some claim that the combined properties of this timber make it better and more appropriate for this type of use than Greenheart! It has uses for bridge building, joinery and flooring, particularly where a high level of pedestrian traffic is expected. It has an occasional use in kitchens for worktops.

Radial

Tangential

Source:

West, Central and East Africa.

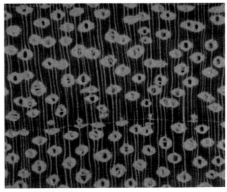

Cross Section

21

AGBA

Gossweilerodendron balsamiferum

Also known as: Achi, Moboron, Noboron, Tola, Tola Branca or White Tola.

Mainly found in the forests of West Africa, especially Nigeria, the Agba tree often reaches heights of well over 100ft (30m). A creamy white wood when first cut, it darkens to a pale yellowy orange, with little differentiation between heart and sapwood. Freshly cut lumber can show slight resin exudation on the surfaces with a tendency for there to be larger isolated pockets of gum. The grain is generally straight with some evidence of interlock occasionally occurring. Its appearance is similar to Mahogany with coarse but evenly scattered pores visible on the face surfaces. The overall impression is of a bright, clean surface that feels silky when planed.

Agba is not an excessively heavy timber at around 14kg (30lb), on average per cubic foot when dry. It is not as strong as Beech, but it does have excellent resistance to crushing strains and is stiff but not brittle. Agba is susceptible to brittleheart, shakes running across the grain, and careful selection of material can be necessary. It dries well, although the resin can cause problems if it gets too hot. I have seen charges coming out of kilns with stalactites of resin attached to the boards! Once dry, this problem disappears. Agba is classified as durable, and there is some resistance to pressure preservative treatment of the heartwood but the sapwood is permeable. Shrinkage is minimal during drying and the timber is generally very stable with little further movement.

The working properties of Agba are good without too much blunting of saws and planes. Some resin build-

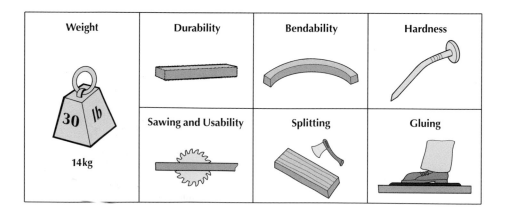

Weight	Durability	Bendability	Hardness
30 lb / 14kg			
	Sawing and Usability	Splitting	Gluing

up can occur when sawing if it is present in the raw material. Nailing and screwing are not a problem providing that care is taken. It glues well when dry, but try to avoid gluing resinous material. Because the grain is quite open, some filling may be necessary before staining and polishing; however, it stains well and finishes in a satisfactory manner.

This is a good general purpose timber that can be used for a wide range of applications including furniture. The natural durability of Agba's heartwood allows for external joinery as well as internal; care needs to be taken to select heartwood if possible. When a reasonably light but strong wood is required Agba may fit the bill, although any brittleheart needs to be avoided. This timber is not suitable where it might come into contact with foodstuffs.

Radial

Tangential

Source:

Predominantly West Africa.

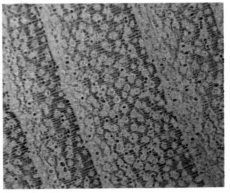

Cross Section

ALDER

Alnus spp.

Also known as: Black, Grey, Oregon or Red Alder.

Black Alder (*A. glutinosa*), and Grey Alder (*A. incana*) are generally found in Northern Europe, the UK and parts of western Siberia and Asia. There is very little to distinguish between the two. They may be marketed together with some reference to White Alder if the wood is consistently pale enough. Alder from North America (*A. rubra*), is sold on the market as Red or Western Alder and occasionally as Oregon Alder. When first cut, Alder is nearly white but darkens to a pale yellow or light reddish brown with a dull surface. There is little to determine the difference between heart and sapwood. The wood is rather dull with no apparent figuring. If cut on the quarter it is possible that some figuring will be revealed, but this is not consistently the case.

Alder weighs on average around 15kg (33lb) per cubic foot when dry and is not particularly strong. It dries well and fairly rapidly without too much distortion or shrinkage. Alder is not durable and if the sapwood is left untreated it will be easily infected with furniture beetle and the like. Watch out for sap stain fungi in wet wood and the residual blue stain that may occur. When immersed completely under water the heartwood will remain sound for a considerable time, hence its use for the piles of Venice.

This wood works well with hand or machine tools due to its lightness and generally straight grain, although if wild grain is present it will pluck out. Keep your cutting edges sharp and fine for best results. It nails and screws well with no need for pre-drilling

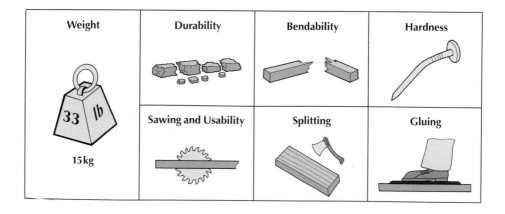

Weight	Durability	Bendability	Hardness
33 lb **15kg**	Sawing and Usability	Splitting	Gluing

except at the end of boards. It is not much use for bending, as it splits and buckles, but is excellent for turning. It glues well, takes stains and can be polished to a good surface finish.

Alder is a rather dull timber and therefore has little use as a show wood, although allegedly it was used to make Stradivarius violins! It has properties that have led to its use in turnery and as a pattern wood. Historically it has been used for making clogs, brush backs, toys and gunpowder charcoal. It can also be peeled and used for plywood. Red Alder is employed as a cabinet wood where figure does not have to be a feature. Its working properties lead to its use in small mouldings. Alder's bark is used for tanning.

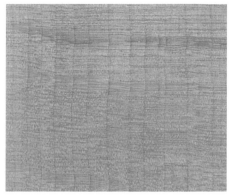

Radial

Tangential

Source:

Predominantly Europe and North America.

Cross Section

ANDIROBA

Carapa guianensis

Also known as: Bastard Mahogany, Carapa, Crabwood, Figueroa, Krappa, or Tangare.

Also known widely by its other main common name, Crabwood, this South American timber is closely related to the Mahoganies but should not be confused with them or called by that name. When freshly cut, Andiroba's colour ranges from pale pink through to dark red with a clearly defined sapwood of pale brown or oatmeal. After exposure and seasoning the heartwood darkens to an even dark reddish brown. The grain is fairly straight with some tendency to interlock occasionally, which produces a broken stripe or mottled figure effect.

Andiroba weighs on average around 18kg (40lb) per cubic foot when dry. It seasons slowly with a tendency to split and warp in the initial stages; however, care needs to be taken not to hurry the process or excessive wastage through degrade will occur. Shrinkage is not great but is probably more noticeable than with some timbers. It is stronger than Mahogany and on a par with Beech. It does not take well to steam-bending. Logs of this timber can be attacked by ambrosia beetles, leaving a distinct pinhole with the possibility of some adjacent staining. Andiroba is rated as moderately durable. It takes preservative well and if used for external joinery should be so treated to avoid any fungal attack.

Andiroba generally saws well with little burning during cross-cutting. The grain can pluck out when planed on the quarter sawn surfaces, or if interlocking is present, but will mostly produce a good, fine finish. If difficulty is found with producing a

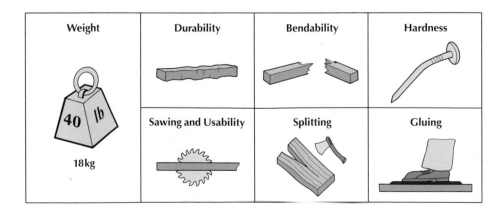

Weight	Durability	Bendability	Hardness
40 *lb* **18kg**			
	Sawing and Usability	Splitting	Gluing

satisfactory finish a reduction in the cutting angle has proved to be the best approach. When nailing or screwing it will be necessary to pre-drill to avoid splitting, especially near the end grain. This wood glues, stains and polishes well.

Andiroba is one of those lesser known species appearing out of South America. At source, it has traditionally been used for general construction, furniture, masts and spars, joinery and some plywood. If available in commercial quantities it could have possibilities for internal mouldings and joinery. For external use it should be treated with preservative and can easily be moulded for window and door components. It is quite useful for furniture. However, the lack of a distinct figure will probably always relegate it to a secondary position in this area.

Radial

Tangential

Source:

South America.

Cross Section

APPLE

Malus sylvestris and *domestica.* (*M. pumila* is Crab Apple.)

Apple is one of the timbers commonly referred to as fruitwood. It is not grown commercially for its wood, so availability is limited. Found throughout the temperate zones, Apple's best properties are brought out when it is grown in colder climates. The result is a close-grained timber with a fine and even texture. Most trees grow irregularly, which produces a rather variable grain pattern. There is little or no definition of sapwood from the heartwood; colour varies from a light pinkish-brown to a reddish hue, similar to Pear.

Apple tends to be fairly heavy at around 20kg (44lb) on average per cubic foot when dry. This results in a strength similar to that of Beech. Extreme care needs to be taken with drying to avoid a high percentage of wastage through distortion. Some pre-drying on thin stickers is advisable prior to artificial drying and a mild schedule is recommended. Shrinkage appears to be limited but little data is found on this subject due to Apple's lack of commercial availability. Movement, once dry, is minimal. Apple is not durable and does not do at all well in variable wet and dry situations. When used in furniture, it is susceptible to attack by the common furniture beetle.

The wood is tough and tends to blunt cutting tools easily, especially when cross-cutting where burning can occur. A reduction of the cutting angle when planing is recommended to minimize the likelihood of poor surface finish due to tearing and plucking out. Some pre-drilling is best towards the end grain but otherwise it nails and screws well. Apple turns well and its fine, even texture lends itself to carving. It will glue well, but care

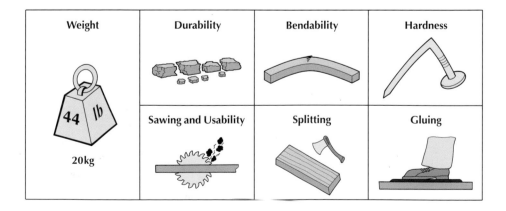

Weight	Durability	Bendability	Hardness
44 lb / 20kg			
	Sawing and Usability	Splitting	Gluing

needs to be taken when joining up a width to alternate the grain direction. It accepts stains and polishes up nicely to produce a smooth finish.

Historically, Apple has been used for tool handles because it polishes naturally with use. It is good for mallet heads and has additionally found a limited application in skittle balls. It has been used extensively in the past for 'country-made' furniture, when it was mixed with other timbers such as Pear and commonly called fruitwood. Old-style cottage doors were often made from Apple wood. Today it is frequently utilized as a craft wood for turning and carving with, additionally, a limited amount of inlay work.

Radial

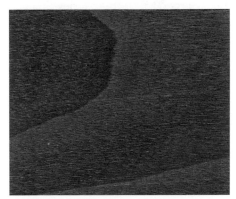

Tangential

Source:
Throughout Europe and similar climates.

Cross Section

ASH, AMERICAN

Fraxinus spp.

Also known as: Black, Brown, Green or White Ash.

White Ash is generally confined to *F. americana*, Green to *F. pennsylvanica* and Black or Brown to *F. nigra*. In the trade White Ash is popular, but often there is little or no differentiation between the species except for Black Ash, which is normally marketed and sold separately. Care needs to be exerted if all 'white' wood is required, as some brown or black coloration can be expected unless otherwise specified. The timber is a creamy white colour with the sapwood showing slightly lighter than the heartwood. Black Ash tends to be a little darker and is not as tough as the others. This is a coarse, open grained timber that can be planed to a smooth surface finish.

It weighs on average around 18kg (40lb) per cubic foot when dry. It is slightly lighter than its European counterpart but has many similar qualities. This is a strong and extremely tough timber, although the Black Ash tends to be slightly less so. These properties have made it very popular for use in tool handles, especially in striking tools such as hammers and axes. It generally dries well but has a tendency to split at the ends when forced in an artificial environment. Distortion is limited, but shrinkage can be above average because of the open-grain structure. Rated as non-durable, American Ash should be preservative-treated for all external applications.

American Ash saws and planes well without too much blunting effect on tooling. The tendency to splitting can be increased when nailing and screwing, so pre-drilling is highly recommended. It steam-bends

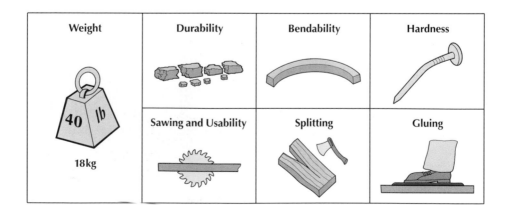

Weight	Durability	Bendability	Hardness
40 lb **18kg**			
	Sawing and Usability	Splitting	Gluing

reasonably well with little tendency to crush or split. It also glues well, with its open grain allowing good penetration. However, staining can be a bit of a problem when trying to achieve an even finish, as the wide, open grain tends to accentuate and absorb stain in differing amounts. Once this has been overcome, a good polished surface can be achieved.

Ash has been used extensively for tool handles, in sports equipment for pool cues, baseball bats, hockey sticks, cricket stumps and tennis rackets. Its bendability and turnability have been exploited to develop it as a material for use in furniture manufacture. Ash is a good, all-round timber that can be used in a wide variety of circumstances where durability is not an issue but toughness is.

Radial

Tangential

Source:

North America.

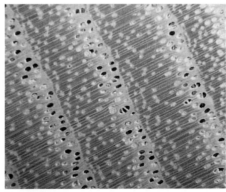

Cross Section

31

ASH, EUROPEAN

Fraxinus excelsior

Depending upon origin it is also known as: English, French and so on, as well as Olive Ash.

Found throughout Europe, Ash is a fairly tall-growing hardwood with a good trunk length before branches occur. In the past, country folk polarded the Ash, cutting off the tree above ground level and then allowing it to spurt from the stump. After a few years, the best and strongest shoot was selected and left to grow on, while the others were removed. This helped to create straight, long trunks as the shoots competed with each other for space. The wood is creamy white in colour with no differentiation between sap and heartwood. Some dark brown or black streaks can be found, hence the term Olive Ash. Flat-sawn surfaces show well-defined grain patterns due to the strong growth rings. The grain is normally fairly straight but coarse in texture.

Ash is a medium-weight timber at around 19kg (42lb) on average per cubic foot when dry, although this can vary widely depending upon growing conditions – slower is heavier and quicker is lighter. Ash is a very tough timber with great elastic values, making it popular for tool handles. Its strength is comparable to that of Beech. It dries well and rapidly with little tendency to split or check, although end splits will get worse during the drying process so it is a good idea to nail a strap across the ends of the boards. During kiln drying the temperature needs controlling to avoid excessive heat and therefore distortion. Shrinkage can be above average, especially in the wider, coarser grained timber. Ash is classed as non-durable. For external use it

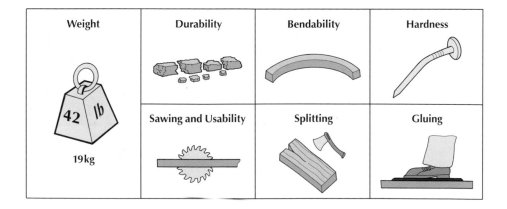

Weight	Durability	Bendability	Hardness
42 lb / 19kg			
	Sawing and Usability	Splitting	Gluing

should be treated and, in the main, will take a preservative reasonably well.

All machining processes are satisfactory with little blunting taking place. Surfaces can be planed to a glass-like finish on dry material. It is a good idea to pre-drill when nailing or screwing because of the tendency to split. Ash steam-bends easily and well, and glues, stains and polishes as per American Ash.

Traditional uses of European Ash include tool handles, oars, spokes and all those already mentioned under American Ash. The ability to steam-bend easily lends Ash to furniture components and to boat-building, where treatment is necessary. A good, all-round wood that can be used wherever a tough timber is needed.

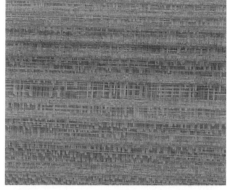

Radial

Tangential

Source:

Throughout Europe and similar climates.

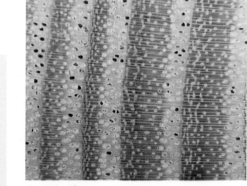

Cross Section

33

BALSA

Ochroma spp. (normally *O.lagopus*).

Also known as: Ceibon lancro, Guano, Lanilla, Lano, Polak or Pau de balsa.

Balsa is found throughout South America and is probably the best known lightweight commercial hardwood. It is a rapid-growing tree, easily capable of attaining 10ft (3m) per year in well drained, rich, moist soils. Depending upon growing conditions, the weight per cubic foot can vary considerably. Some records indicate anything from as little as 3kg (7lb) up to over 9kg (20lb) per cubic foot. In some instances, this weight variation can occur in the same tree! Balsa is white in colour with the occasional pink or brown tint, while the central core can be dark brown. The grain is of a coarse, open texture, normally straight, and has a silky lustre when freshly finished.

For commercial use, some selection for weight uniformity takes place, with the range generally being from 2–5kg (4–11lb) per cubic foot when dry. Balsa's lightness in weight also provides good sound and heat insulating properties. Naturally, it is not a strong timber when compared to other woods, although its weight to strength ratio is quite good. To avoid excessive splitting and distortion Balsa is generally exported green and kiln-dried in the importer's yard. It is a difficult timber to dry, with warping and splitting being the main problems. Case-hardening is often apparent and in some instances the outer skin can have been so overheated that the timber appears 'toasted'. Considerable shrinkage also takes place from green to dry.

Balsa will crumble when being worked if thin, sharp tools are not used. It takes nails and screws well but is not really strong enough to retain

Weight	Durability	Bendability	Hardness
20 lb 7 or 9kg			
	Sawing and Usability	Splitting	Gluing

them, so gluing is the best way to joint this timber. Naturally, it is not durable and needs to be treated if used in exposed conditions. The openness and lightness of the material will lead to some leaching of preservatives near the surface. It will take stain, by the gallon, and polishes well with care.

Traditionally, Balsa is well known for its use in model-making, but some of its properties find commercial applications. For example, it is an excellent insulator for refrigeration in cold stores, and its lightness has led to uses as floats, buoys, rafts and lifebelts when preservative-treated. In the Second World War it was used for parts in aircraft manufacture. Balsa's softness also lends for use as a packing material for delicate instruments, and so on.

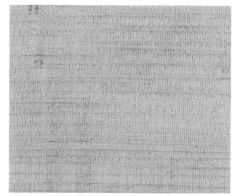

Radial

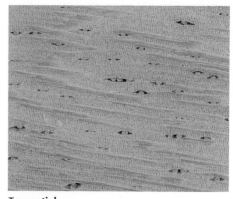

Tangential

Source:

South America.

Cross Section

BASSWOOD

Tillia americana

Also known as: American Lime, or American Linn in the UK, Linden or Whitewood. (Basswood is also a vernacular name applied to *Endospermum spp.*, a Far Eastern tropical timber better known in the trade as Sesendok).

Basswood is a North American hardwood similar in appearance to European Lime, hence the use in the UK of the name American Lime. The wide sapwood is of a creamy, nearly white colour and merges into a light yellowish brown heartwood that may occasionally have some darker streaks. It has a fairly close, straight and evenly textured grain, generally without distinct markings. When freshly cut it has a pleasant smell, but this disappears after it has dried.

Weighing on average around 11 to 12kg (24 to 26lb) per cubic foot when dry, Basswood is a lightweight. It is not a strong timber, it does not bend well and has little resistance to hard impact. It dries readily with only a small amount of distortion or splitting and shrinkage is reasonable. Once dried it will remain stable with little or no movement. Basswood is not durable and should be treated if used in exposed locations; it takes preservatives readily.

Basswood's working properties are good. It saws well without too much blunting and can be planed to a clean, smooth finish. The fine, even grain texture coupled with its general stability and ability to hold an edge make it an ideal timber to carve. When nailed it has good resistance to splitting, so there should be little need to pre-drill except near the end grain. It also screws well and readily takes glue. Its light weight allows staining to be easily achieved,

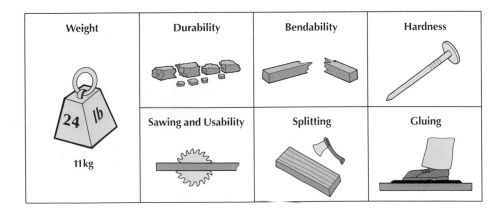

Weight	Durability	Bendability	Hardness
24 lb / 11kg			
	Sawing and Usability	Splitting	Gluing

although some darkening near the end grain through greater absorbency needs to be watched. It will polish to a fine finish but has a tendency to bruise if knocked.

Basswood is not as popular as European Lime for carving and pattern work, although it does have many similar properties and has been successfully utilized for this purpose. Its turning properties are excellent. It is often used in America for plywood corestock, edge bandings, venetian blinds and door manufacture. Its lack of odour has lent this wood for use in food containers and preparation boards. It can also be employed for picture framing, parts for musical instruments and piano keys, and in small, decorative mouldings.

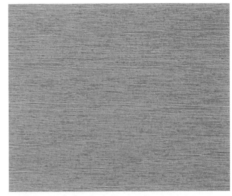

Radial

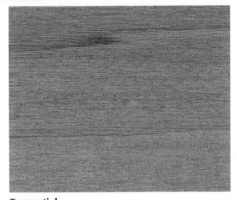

Tangential

Source:

North America.

Cross Section

BEECH, EUROPEAN

Fagus sylvatica

Depending upon origin it is known as Danish, English, French, German, Romanian, and so on.

Beech is found throughout Europe. It is a white-coloured wood when freshly cut, but it may have a tendency to be pale brown with pinkish or dark brown streaking. The steaming of Beech is a process adopted to make this coloration more uniform, resulting in a deeper pinkish brown. This practice has generally been carried out in Romania and the former Yugoslavia. There is little distinction between heart and sapwood. Beech can fairly easily be identified by the tiny flecks produced by the ends of rays on the tangential section.

In general fairly hard and dense, when dry Beech weighs on average around 20kg (44lb) per cubic foot. It is a medium strength timber and is used regularly as a benchmark against which to measure others. It dries fairly well and rapidly, but much shrinkage occurs. It is also slightly refractory and tends to distort, split and check. Once dry, little further movement will take place so long as it is kept in fairly constant conditions. This timber is not durable and is rarely used in exposed locations.

Beech saws and planes fairly well but with some tendency to burn if tools are not kept sharp. When bandsawing large amounts of trailing, breakout can occur on the bottom of the cut. If the timber has been case-hardened during drying the release of that tension when cut can cause considerable distortion and movement. Pre-drilling is necessary when nailing, especially near edges, and always pre-drill when screwing. Beech glues well but avoid getting any

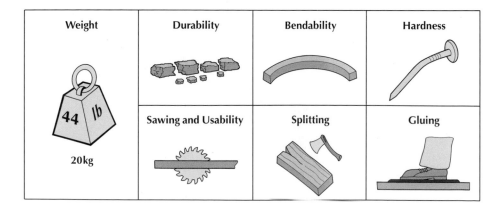

Weight	Durability	Bendability	Hardness
44 lb / 20kg			
	Sawing and Usability	Splitting	Gluing

on surfaces that might be stained. It will stain to match just about any other timber and polishes to a fine finish. One of the great things about Beech is that it will steam-bend with ease, but watch out for pin knots as it is likely to break at these points.

Beech is a versatile timber and is used extensively for internal applications. It is a traditional material for chair manufacture as its bendability makes it suitable for various necessary components. It turns well and is easily shaped and carved. Beech-faced ply is used for panels and infill work; it can also be moulded and shaped around various contours. Its hardness and density lead to use as handles for tools, brushes and as a good flooring material made very popular by the Danes.

Radial

Tangential

Source:
Throughout Europe and similar climates.

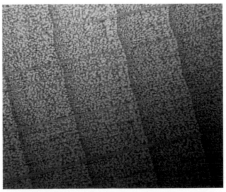
Cross Section

BINTANGOR

Calophyllum spp.

Also known as: Bunut, Mentangol or Penagayyer.

Calophyllum as a genus is found throughout the tropics; it is especially prevalent in South-east Asia. Bintangor is a grouping of those trees that are available in reasonable commercial quantities, and it is generally sourced from Malaysia. It is not to be confused with other *Calophyllum* from Papua New Guinea and adjacent areas that are marketed commercially under the generic name of Calophyllum.

Bintangor's sapwood is a pale pink-brown and is clearly discernible from the heartwood, which has a striking open reddish brown grain; this darkens on exposure and use. It has a coarse texture with an uneven grain that is sometimes interlocked or wavy. Some quarter-sawn material will occasionally display a stripy figure and there is often evidence of dark zigzag streaking on the tangential face. When planed, Bintangor has a smooth, lustrous finish, although overall the timber is not particularly attractive

Not heavy, Bintangor weighs around 18 to 19kg (40 to 42lb) on average per cubic foot when dry. Care needs to be taken when drying because of its tendency to warp. Extra stickers are recommended and weights on top of the load will help. However, apart from the distortion, degrade during drying is minimal. Bintangor is not particularly strong and should not be considered for use in structural applications. It is non-durable timber and is best used internally unless treated with preservative, which it takes well.

This timber will saw well so long as no tensions have been created during

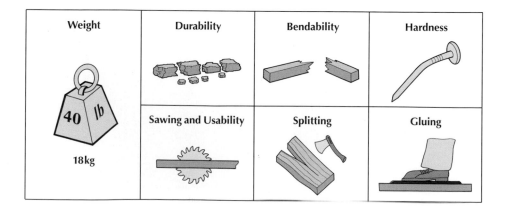

Weight	Durability	Bendability	Hardness
40 lb / 18kg			
	Sawing and Usability	Splitting	Gluing

the drying process. If case-hardened, even slightly, it is likely that a lot of reject will occur when cut. A good way to avoid this is to cut to the nominal size prior to drying. It will plane well and produces a smooth finish; pre-drilling when nailing or screwing is necessary. Bintangor's open grain texture allows good penetration of stain, but avoid darkening near the end grain. It glues well and will polish to a reasonable finish when the grain is filled, but it is not a decorative timber.

Radial

This is a general utility timber used for a wide variety of purposes at source. Although not particularly durable, it has some light constructional applications. Furniture and components are occasionally made from Bintangor. It has apparently proved to be excellent for swimming pool diving boards.

Tangential

Source:

South East Asia.

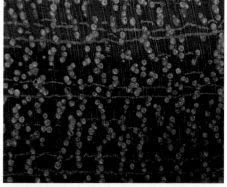

Cross Section

BIRCH, EUROPEAN

Betula spp.

Depending upon origin it is known as English, Finnish, Swedish, and so on.

Found throughout Europe, Birch is probably one of our best known timbers. A hardy tree, it is most commonly seen in Northern and Eastern Europe and survives right up to the polar regions. In Scandinavia, where it grows in stands, the boles are often clear and straight. It is from these that the famous Finnish Birch ply is produced. The wood varies from white to a light pale brown with little or no distinction between heart and sapwood. It has no definite figure or grain pattern, but does have some slight silvering and darkening that is easily recognised.

Birch is a fairly lightweight timber at around 18kg (40lb) on average per cubic foot when dry. It is fairly tough when dry, similar to Ash, and comparable to Beech in strength. It must not be left in a freshly sawn state for too long or fungal growth will appear on the surface. It dries fairly rapidly but has a tendency to distort; it is useful therefore to weight the top layers down if possible. Shrinkage is quite high from green to dry, but if it is kept in stable conditions thereafter little further movement should occur. Birch is perishable and is therefore strongly recommended for internal use only, taking a preservative well if necessary.

Birch saws reasonably well and generally a good planed surface can also be achieved. On occasions, the finished surface will need further attention to remove woolly areas; some tear can occur around knots. It can be bent successfully if the knots are avoided, although that may prove difficult unless the wood is from a clear Scandinavian source. Pre-drill

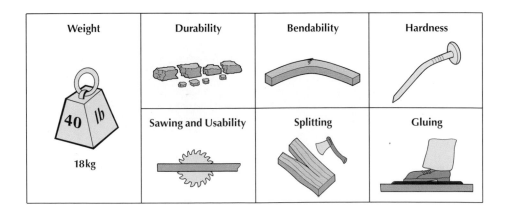

Weight	Durability	Bendability	Hardness
40 lb / 18kg			
	Sawing and Usability	Splitting	Gluing

when screwing and especially when nailing near edges. Birch peels well and is used extensively for plywood. It stains and glues well, and the finished product can be polished satisfactorily.

Birch is probably best known for its Finnish Birch plywood, which is used extensively for furniture and other cabinet work. The poorer quality material has also been used traditionally in furniture for framing in areas where a few knots and splits are acceptable. It turns quite well and therefore has a use in brushes and brooms. It is often used as pulpwood in paper production. Interestingly, beer can be made from the sap and the bark can be distilled to provide oils containing methyl saliyate, the main component of aspirin.

Radial

Tangential

Source:

Throughout Europe and similar climates.

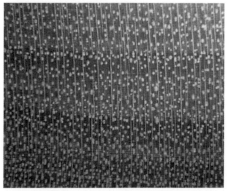

Cross Section

BOXWOOD, EUROPEAN

Buxus sempervirens

Also known as: Box, Iranian, Persian, Turkish depending upon origin.

As a trade name, Boxwood covers various timbers that are not truly related species. This description relates to European Boxwood, but other woods with similar appearance and nature may be confused with it. Boxwood is found in various locations throughout Europe and into Asia Minor. Because it never grows much past the size of a small tree or bush, planks and board sizes are not of large dimensions; nevertheless, fairly consistent quantities are usually available commercially. It is pale or light yellow in colour, with no apparent demarcation between the heart and sapwoods. The grain is usually tight, straight and of a very even texture. Smaller trees will tend to have poorer quality grain that may be quite irregular.

It is a fairly heavy timber, varying from around 25 to 26kg (55 to 57lb) on average per cubic foot when dry. This has a bearing on drying times, which are slow and can lead to excessive surface checking. It is understood that some pre-treatment prior to drying can take place to alleviate this by soaking the lumber in a mixture of salt and urea. Advice needs to be taken over this and consideration given to the end use, as this may be affected by these chemicals. Boxwood is considered to be a very hard and strong timber. It is rated as durable, but because of its size it is unlikely to be used for external applications.

Sawing Boxwood can at times be difficult depending upon the grain structure. Firm pressure is required when planing to avoid chatter and any irregular grain will pluck out. Care

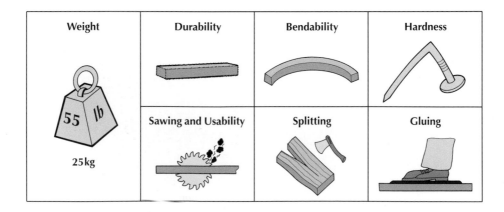

Weight	Durability	Bendability	Hardness
55 lb / 25kg			
	Sawing and Usability	Splitting	Gluing

need to be taken with all cutting tools, so keep the edges sharp. Boxwood turns well and produces a fine finish. If nailing or screwing, pre-drill to avoid splitting. It glues well, stains are readily accepted and the grain lends itself to a fine polished finish.

Because of its tight, even grain texture and its ability to hold detail, Boxwood has been used for generations as an engraving block. Its properties lend themselves to use as scales, rules, inlay, parts of musical instruments and many other similar applications. Its good turning properties have ensured a use as tool handles, chess pieces, rollers, and so on. In the textile industry, it is used to make shuttles and specialist rollers, especially in silk manufacture.

Radial

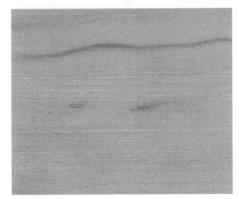

Tangential

Source:
Throughout Europe and similar climates.

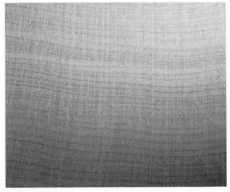

Cross Section

CAMBARA

Erisma uncinatum

Also known as: Cedrinho, Felli Kouali, Manonti Kouali, Mureillo, Quarubarana, Quarubatinga, Quaruba Vermelha or Singri-Kwari.

Sourced mainly from Brazil, Cambara is the most popular name in use by importing countries. It is found throughout South America, and as can be seen has a variety of other local names. The heartwood is a light pink-brown through to a darkish purple-brown with a distinct pale white sapwood. The grain texture is open and fairly straight. It has no major distinguishing features, being rather plain but consistent in appearance.

This is a light- to medium-weight timber, averaging around 20kg (44lb) per cubic foot when dry. Kiln-drying Cambara can be troublesome, especially the heavier timbers. Thinner sizes are most successful; boards over 5cm (2in) thick can be

difficult. With thicker material kilning can produce a high percentage of reject and can be very protracted, although long periods of pre-drying may help to avoid this. There is some tendency to move after kilning if care is not taken. Shrinkage is not particularly significant, although tangentially it may be up to eight or nine per cent. Some surface checking can occur. Its strength properties are probably just slightly less than those of Beech. Cambara is classed as moderately durable. It appears to be reasonably amenable to preservative treatment and should be treated if used in external joinery.

Interlocked or spiral grain is rare, so sawing Cambara is relatively easy. When planing, standard or high speed steel tipped blades produce reasonable finishes without too much blunting or tear out. Cambara will nail

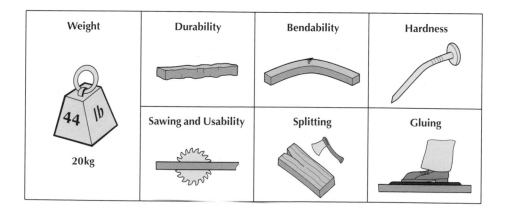

Weight	Durability	Bendability	Hardness
44 lb 20kg			
	Sawing and Usability	Splitting	Gluing

and screw relatively easily, but as with most other timbers it is advised that pre-drilling takes place near the end grain. It can be peeled or sliced readily with only a slight blunting of tools. Gluing is satisfactory. It is amenable to stains and polishes reasonably well.

Cambara is one of those slightly lesser known species that is not always available in large commercial quantities, although it is often considered as an alternative to some of the Far Eastern and African hardwoods currently used for general joinery purposes. It has potential for furniture. It has been used for panelling, plywood and packaging. Its good planing properties lend themselves to interior mouldings of all descriptions, so long as no case hardening is present in the stock material. A good all rounder.

Source:

South America.

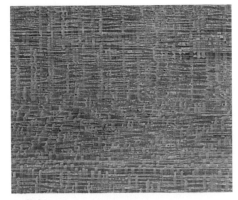

Radial

Tangential

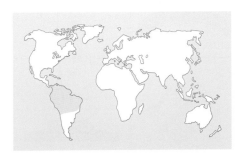

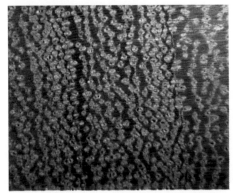

Cross Section

CHERRY, AMERICAN

Prunus serotina

Also known as: American Black Cherry, Rum Cherry, Wild Black Cherry or Wild Cherry.

This is the only *Prunus* of commercial value found across North America and is very similar in colour and texture to its European cousin. Its heartwood varies from a warm light to dark reddish brown and has a nice lustre when planed. The sapwood is a distinctive creamy white and is fairly narrow in old, slower grown trees. Its overall grain features are not dramatic and are similar to those of Birch, with a darkening that follows the growth patterns and some occasional flecking.

A medium to lightweight timber averaging 16 to 17kg (35 to 37lb) per cubic foot when dry. This Cherry dries fairly rapidly but can distort; end splitting is another problem that occurs, but once dry there is little further shrinkage or movement. Slightly lower in strength than Beech, it is not often that Cherry will be used for structural purposes, and despite being classed as moderately durable, it is unlikely to be used for anything other than internal purposes. If used externally, treatment is recommended. The sapwood is susceptible to attack by wood borers and should be avoided in furniture.

American Cherry saws well so long as the tensions set up during drying are not too great. Because it has a fairly uniform grain texture it has good machining properties, planing up to a smooth, lustrous finish. Some plucking out of the finished surface can occur in an irregular-grained material. It can be peeled and is used quite extensively for decorative veneers. Nailing and screwing are good, but always pre-drill to avoid

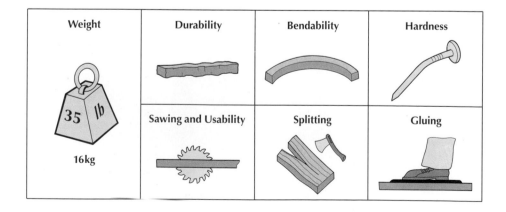

Weight	Durability	Bendability	Hardness
35 lb 16kg			
	Sawing and Usability	Splitting	Gluing

unnecessary splitting. It bends well and can be steamed into various furniture components with reasonable ease. Cherry glues and takes stain well, but is normally used in its natural state. The fine-textured grain lends itself to an excellent finish when polishing.

American Cherry is readily available in commercial quantities, making it a preferred choice over its European cousin. It has a mass of uses for internal joinery and furniture. Stable once dry, it is used principally for fine furniture, cabinet-making, panelling and other decorative joinery. It has also found some limited use in gun stock manufacture and for musical instrument components. Mass-produced kitchen and bedroom furniture doors are often to be found in Cherry. Its bark is used to flavour rum and brandy.

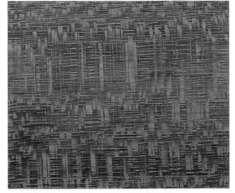

Radial

Tangential

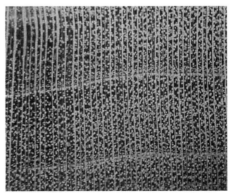

Cross Section

Source:

North America.

CHERRY, EUROPEAN

Prunus avium

Also known as: Gean, Mazzard or Wild Cherry.

Unlike the American variety, this Cherry is not renowned for its commercial availability. It is grown throughout Europe and Asia Minor and has a reputation for blossom rather that as a timber resource. The best wood will come from old forest growth. Trees from orchards are unlikely to produce good enough boles to convert into lumber; yields are likely to be small, therefore adding to the overall basic cost of the wood. The colour varies from a pale pinkish to light golden brown with a hint of green. Freshly sawn timber has a pleasant smell that is often compared to a rose blossom fragrance. In antique furniture it is one of the timbers often lumped together with others and referred to as fruitwood.

A light to medium weight timber at around 17 to 18kg (37 to 40lb) on average per cubic foot dried. It will dry fairly rapidly, but irregular-grained material distorts extensively. End splits are common and cleats should be fitted or the ends sealed to help overcome this. Shrinkage is average and once dry Cherry should have little further movement. A light gum-like substance can be apparent on the surface but this is easily removed if crystallized; some oils can remain in the pores. It is slightly weaker than Beech. Like its American cousin, European Cherry is not used for structural purposes. Nor is it particularly durable, and should be treated if used in exposed positions.

Cherry saws well, but has a tendency to burn or distort further if the wood contains any irregular grain. Straight-grained timber planes well

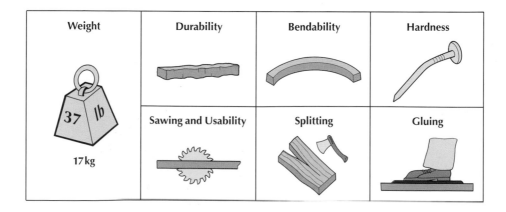

Weight	Durability	Bendability	Hardness
37 lb / 17kg			
	Sawing and Usability	Splitting	Gluing

with little blunting, but some other grain patterns will pick up. Pre-drill all fixings to avoid splitting, especially near the end grain. Cherry can be bent and steams well. One European practice is to steam in a similar fashion to Beech to produce a more uniform, darker pinkish red colour. It glues well and takes stains fairly readily, although because oils can be found care needs to be taken in some instances when polishing.

This is a decorative wood that has found its way into many fine pieces of furniture and cabinet-making. Its tendency to warp sometimes leads it to be used in smaller sections. Some solid wood and radial sliced veneers are used in panelling, and it is often found in old cottage doors mixed with other fruitwoods. Cherry turns well and is used for fancy knobs and similar applications.

Radial

Tangential

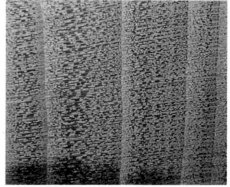
Cross Section

Source:
Throughout Europe and similar climates.

51

CHESTNUT, HORSE

Aesculus spp.

Also known as: Yellow or Ohio Buckeye in North America.

The Horse Chestnut tree grows throughout the northern temperate zones, mainly in the warmer regions. Most commonly known in the UK as the conker tree, Horse Chestnut is not usually found in large commercial quantities. In spring, the tree has flower spikes of cream or bright pink; it is often seen at the side of the road or in a parkland setting. In autumn, the fruit is contained in a green, knobbly outer shell that when opened reveals a leathery cased nut. This nut is not edible but makes a great conker for the kids to play with!

The creamy grey-white sapwood of this Chestnut merges without any definite demarcation with the heartwood, which tends to be creamy or straw white. The grain is fairly open and is usually straight; it has an even, uniform texture with the exception of older, stunted or misshapen trees that may be irregular.

This is not a heavy wood, with a dry weight average of around 14kg (30lb) per cubic foot. It dries well and fairly rapidly without too much distortion in the straighter grained material; shrinkage is only moderate. Once dry and in favourable conditions it remains fairly stable. However, Horse Chestnut is not a particularly strong timber and can tend to be brittle at times. It can be steam-bent, although care needs to be exercised. It is not durable and needs preservative treatment, which it takes readily, for any outside applications.

Machining this timber can be difficult if blades and tools are not kept sharp; otherwise, its slightly woolly nature leads to poor surface finishes. It is easily and often turned

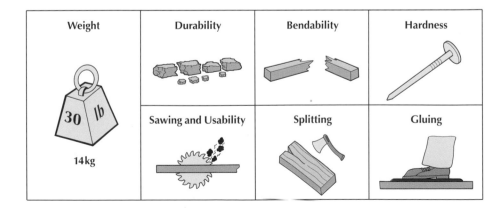

Weight	Durability	Bendability	Hardness
30 lb / 14kg			
	Sawing and Usability	Splitting	Gluing

for small utility items on copy lathes where it performs reasonably well. As a light-weight wood it nails relatively easily without too much splitting; pre-drill for jointing with screws. Its open grain aids gluing functions, but this will need filling before staining and polishing.

Often batched and used with other lesser known or exploited commercial species, Horse Chestnut is not particularly valued for its wood. Under these circumstances it will probably be used for pallets, dunnage or packing case manufacture. Straight-grained material is suitable for engineering pattern-making due to its fairly stable nature. It is often used for small turned kitchen or utility items and brush backs. It is not commonly seen in furniture manufacture, but should not be dismissed because of restricted supplies.

Radial

Tangential

Source:

UK and North America.

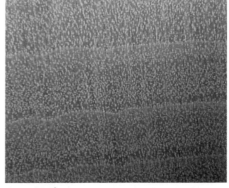

Cross Section

CHESTNUT, SWEET

Castanea sativa

Also known as: European or Spanish Chestnut.

This timber is not to be confused with the Horse Chestnut, *Aesculus hippocastanum,* or conker tree found in the UK. Sweet Chestnut produces a popular edible fruit. Often sold by street vendors in winter, these roasted chestnuts are delicious!

This tree grows throughout Europe, especially in the south, and on through to Asia Minor and some parts of North Africa. It has an open, course grain that bears some resemblance to Oak, but lacks the prominent medullary rays found in that wood. The heartwood varies from a medium yellowish to brown, while the sapwood is a distinct cream colour. Old individual trees may have some spiral grain, but in general it is fairly straight. Like Oak, it has a slightly acidic content that can corrode metals and cause staining, especially in wet wood.

Another light to medium weight wood that averages 15 to 16kg (33 to 35lb) per cubic foot when dry. It can be difficult to dry with some evidence of collapse or honeycombing, and it also has a tendency to retain moisture in patches. Remedial treatment of these defects is difficult. Shrinkage and movement are not major problems, although the timber should be thoroughly dry before use. It is not a particularly strong timber and therefore should not be used in large structural projects. The heartwood of this timber is fortunately classified as durable because it does not take preservative well!

Sweet Chestnut is fairly easy to work, sawing and planing well. It is not a particularly good bender, with some rupturing and wrinkling

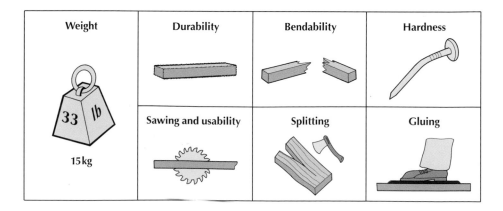

Weight	Durability	Bendability	Hardness
33 lb / 15kg			
	Sawing and usability	Splitting	Gluing

occuring on the inner faces. It nails and screws well, but requires pre-drilling near the end grain. Once planed, it will take stain and glue well. Some filling is necessary to produce a smooth surface prior to polishing.

This timber has a wide range of uses both internally and externally. The tree can be coppiced to produce poles for fencing. The poles are split rather than sawn and it is valued for this use because of its natural durability. Its likeness to Oak has lead to a widespread use as a substitute, especially in furniture. Although not as attractive as Oak, not many can tell the difference. Traditional uses are for ladder rungs, barrel staves, gates, coffin boards, agricultural implements and many similar items.

Radial

Tangential

Source:

Throughout Europe and similar climates.

Cross Section

55

COCOBOLO

Dalbergia retusa

Also known as: Granadillo, Nambar or Palo Sandro.

This tree is of medium size, probably growing no more than about 30m (100ft) high. With a fluted trunk, bole sizes do not exceed more than 60cm (2ft) maximum in diameter. Cocobolo is found throughout Central America, predominantly on the Pacific coast side. It is not available in large commercial quantities and is sourced, in billets, mainly from Mexico, Costa Rica and Nicaragua. Cocobolo is related to the Rosewoods and has many similar features but is much heavier and harder. The wood has a slight fragrance when cut and feels cold to the touch. The sapwood is a dirty cream or grey colour and is clearly visible. However, the heartwood is stunning! With a rich range of rainbow colours initially, it darkens to a mix of dark golden browns, oranges and deep reds with darker streaks. The wood is of a fine, even and uniform texture with a fairly open and coarse grain.

This is a heavy and hard wood. It will average, when dry, 32kg (70lb) per cubic foot. This is very nearly the same weight as Greenheart! Great care needs to be exercised when drying Cocobolo. If it is too rapid then distortion, checking and splitting will degrade most of the material. Some pre-drying on thin stickers is advisable. As this wood is unlikely to be used in large sizes, conversion initially into thinner and narrower pieces may also help to overcome some of the drying problems. When kiln-dried, it also has a tendency to case-harden. When dried to its final moisture content it is likely to remain fairly stable. This

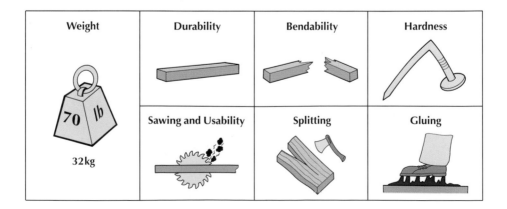

Weight	Durability	Bendability	Hardness
70 lb / 32kg			
	Sawing and Usability	Splitting	Gluing

wood is naturally very durable but is much too expensive to be used for external applications!

The working qualities of Cocobolo are decent, although because of its hardness some blunting will take place. When planing, the wood needs to be held firmly in order to avoid chatter. It is an excellent wood to turn with sharp tools, producing a nice clean finish. Its gluing performance can be variable with care needed. It is unlikely to be stained, and will polish to an exceptionally fine finish.

This is a very attractive and beautiful timber with something of a novelty value, although that may not be reflected in the price. It is used for fancy turnery, cabinet work, rosaries and buttons. It has been particularly sought after for knife handles as repeated washing does not remove the colours.

Radial

Tangential

Source:

Central America.

Cross Section

57

DAHOMA

Piptadeniastrum africanum (Formerly: *Piptadenia africana.*)

Also known as: Agboin, Atui, Banzu, Bokungu, Dabema, Ekhimi, Mpewere, Musese, Singa, Tom or Toun.

Dahoma is found extensively throughout West Africa and less so in the East and South. A large tree, the sapwood is clearly apparent and can be identified as a lighter, greyer hue. The heartwood varies in colour from a yellowish, golden brown to a somewhat drab greyish brown; it can at times look similar to Iroko. It is fairly coarse textured with an interlocking grain that can produce a strip effect on radial-cut surfaces. When freshly cut, it gives off a very unpleasant ammonia-like smell that can cause some irritation; the smell soon disappears once the timber has dried.

A medium-weight timber, Dahoma weighs an average of 18 to 19kg (40 to 42lb) per cubic foot when dry. It does not dry quickly and has variable results. Some distortion and collapse, leading to a washboard-like surface, occurs which cannot be reversed. Best practice advises that the wood is thoroughly air-dried before final artificial drying to avoid as much downgrade as possible. Shrinkage can also be considerable. Dahoma compares favourably with Beech for strength, but it should not be used for gymnasium equipment or ladder rungs in case any of the interlocked grain gives way. The heartwood is classified as durable and is resistant to preservative treatment.

Sawing can be difficult if drying defects are present, and some binding, burning or springing may take place. Its interlocked grain makes Dahoma a fairly difficult timber to plane, with some plucking out of the surface. It can also produce

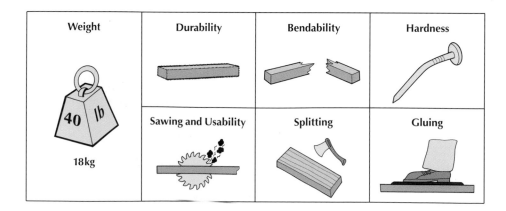

Weight	Durability	Bendability	Hardness
40 lb / 18kg			
	Sawing and Usability	Splitting	Gluing

a woolly, fibrous finish on occasions. Tools should be kept sharp to minimize problems and, if necessary, reductions of the cutting angle should be made to produce a suitable finish. When nailing or screwing, pre-drill if working near the end grain. Generally, Dahoma glues fairly well. To produce a fine surface finish, the grain will need to be filled prior to staining, sealing and finally polishing.

Dahoma is generally used at source with sporadic availability for export. It is useful as a structural timber for lorry bearers and flooring, some harbour and marine works plus general joinery products. It can be peeled and used in plywood production, but tends to split when exposed to the weather. It can be used as a substitute for Iroko, one of the more commonly available hardwoods from Africa.

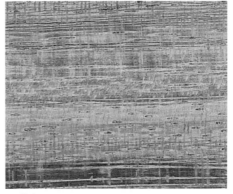

Radial

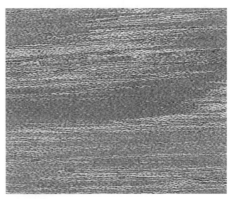

Tangential

Source:

Predominantly West Africa.

Cross Section

DANTA

Nesogordonia papaverifera

Also known as: Apru, Epro, Kotibe, Olborbora, Otutu, Ovoue or Tsanya.

This moderately heavy wood is found mainly along the West Coast of Africa. The heartwood is a dark reddish brown with a clearly visible pale brown to slightly pinkish sapwood. The grain is closely interlocked, often producing a stripe on the radial face similar to Sapele but generally a lot narrower. The texture is fine and even, with occasionally a slight oily or greasy feel when planed. In some instances there can be evidence of small pin-like knots or darker, damaged tissue.

Danta, when dry, weighs around 21kg (46lb) per cubic foot on average. Drying should be undertaken fairly slowly, preferably with some air drying prior to kiln drying. This will help to avoid case-hardening, which can be a feature of this wood. It has a slight tendency to warp and splitting may occur around knots, although shrinkage is not a major problem. In strength, Danta is easily comparable with Beech and has some of the more elastic qualities associated with Ash, but it is not so impact-resistant. The heartwood is classified as durable and the timber as a whole is not easy to treat with preservative. Some evidence of wood borer attack may be seen, usually in the sapwood.

Some slight blunting of tools may occur when sawing and planing this material. The presence of interlocked grain may lead to some plucking out or picking up of the finished surface, especially on the radial face. Tools should be kept sharp and feed speeds adjusted accordingly. All fixings will need to be pre-drilled to avoid splitting or breaking off in the case of screws. On those occasions when an

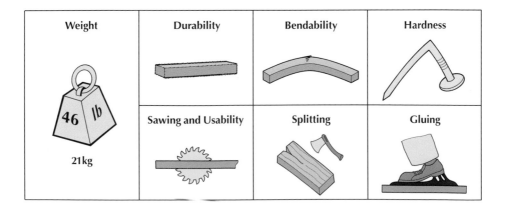

oily finish is evident there may be a problem with gluing and polishing, so some surface treatment may be necessary. It will stain and, with care, fine polished finishes can be achieved.

Danta is a hard-wearing, fairly heavy timber, which is reflected in its uses. It has been used extensively for commercial vehicle construction in both flooring and bodyworks. Its resistance to wear and abrasion has found it used for both industrial and domestic flooring. Its colour and durability have made it popular for internal and external joinery, and it is also used for veneers and plywood. Danta is a timber that should not be ignored when available. With a little effort it can finish well and could easily be used for furniture and similar applications.

Source:

Predominantly West Africa.

Radial

Tangential

Cross Section

DURIAN

Durio spp. (normally *D. zibethinus*)

A number of the *Bombacaceae* genera reach commercial timber size, of which *Durio* is probably the main one. They are spread throughout South-east Asia and are collectively known as Durian. The name is derived from the durian fruit, which is prized and sought after by many for its flavour but not for its smell! The heartwood can range in colour from a pinkish brown through to an orange-brown or deep red-brown. The sapwood is clearly differentiated by a lighter colour. On flat-sawn material the grain is fairly open and coarse in texture with little obvious figure. Some interlocking grain is apparent, with dark deposits in the vessels being a common feature. Quarter-sawn material may well have a ripple effect visible from the ray configuration.

As one would expect, the weight of Durian can vary quite a lot because of the mixed species involved. On average, it is likely to be around 17 to 18kg (37 to 40lb) per cubic foot when dry. It is neither particularly heavy nor hard, and is not a strong timber. Care needs to be exercised when drying. Some species tend to cup badly during the pre-drying stage. The heavier timbers are less likely than the lighter ones to shrink. Within individual parcels, some of the timber will be durable, but Durian is normally classified as non-durable or perishable. It will, however, take preservatives readily, thereby increasing the number of applications for which it can be used.

Sawing and planing Durian can be unpredictable, as some examples will perform adequately in both instances while others will not. In the worst instances, tearing out and burning can occur while sawing. Surface

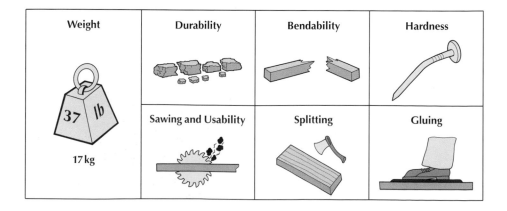

Weight	Durability	Bendability	Hardness
37 lb / 17 kg			
	Sawing and Usability	Splitting	Gluing

finishing can also be difficult, especially on radial faces which can be rough and hard to finish smoothly. Resistance to splitting whilst nailing is good, but pre-drilling for screwing is generally advisable. The timber performs reasonably well during gluing, staining and polishing.

Durian's lack of durability has led it to have fairly restricted uses. It has mainly been used on the home market, with the occasional commercial quantity becoming available for export. It is suitable for light internal joinery such as doors and windows and has been used for flooring and furniture parts. The locals use it for tea chests and packing cases amongst other things. It is not a particularly exciting timber, but one that should be used when available.

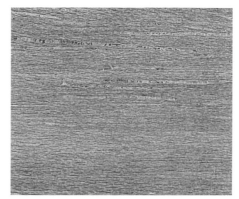

Radial

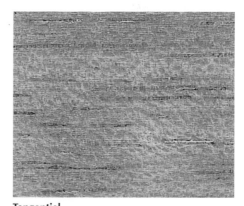

Tangential

Source:

South-east Asia.

Cross Section

EBONY

Diospyros spp. (normally *D. ebenum*)

Ebony may be prefaced with its source country or region, such as Cameroon, Indian, Macassar, Nigerian, and so on, and additionally is known as Adaman Marblewood or Zebrawood.

Ebony is noted and prized for its jet black heartwood, although in reality there are few of the species that give a truly consistent colour. Most have bold, irregular strips varying from bright brown through to a greyish green colour, hence the names Marblewood and Zebrawood. These latter names should not be confused with the South American Zebrawood (*Astronium fraxinifolium*) or the Zebrano from Africa, (*Microberlinia brazzavillensis*). The Ebony tree has a broad, light coloured sapwood that clearly contrasts with the central core of heartwood. In some cases there can be little or no dark coloured heartwood in this central core. The grain can have a straight, or wavy irregular pattern, with a fine and even texture that provides the base for an excellent polishing surface.

There can be some wild variations in the weight of Ebony. The darker, heavier timbers from Sri Lanka will probably be around 29 to 32kg (64 to 70lb) per cubic foot when dry; by comparison, lighter weight timbers may weigh as little as 21kg (46lb). However, whatever the weight, it is still a heavy and hard timber, although these characteristics are not translated into strength because the heartwood can tend to be brittle. Fortunately, it is seldom used in situations where strength is important. Surface checking is a problem when drying Ebony. It should be cut and stacked on thin stickers for some considerable time

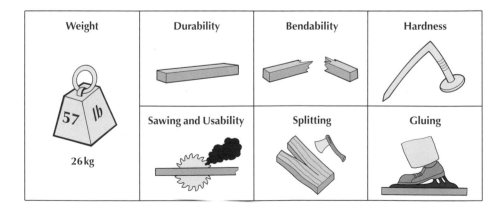

Weight	Durability	Bendability	Hardness
57 lb / 26 kg			
	Sawing and Usability	Splitting	Gluing

before kiln drying. To help avoid this, trees can be girdled before felling, a process in which the bark is cut away right around the trunk, thereby effectively drying *in situ*. The heartwood of Ebony is naturally very durable, so little treatment is necessary.

Because it is so hard and brittle Ebony is difficult to work. Blunting of tools occurs quickly and regular sharpening must be undertaken. Surface finishing can be difficult; reduced cutting angles are found to be the best approach. Pre-drill when nailing and screwing to avoid splits.

Ebony has many decorative uses and is particularly well known as parts for stringed instruments. It can be turned for various items such as chessmen and has been used extensively in cabinet work for inlay.

Radial

Tangential

Source:

Africa, East India and Indonesia.

Cross Section

65

EKKI

Lophira alata

Also known as: Akoura, Azobe, Bongossi, Eba, Hendui, Ironpost, Kaku or Red Ironwood.

This African timber is hard and heavy, with a heartwood that is dark red or chocolate brown and contains some occasional white deposits in the pores. The sapwood is pale pinkish and can be clearly defined from the heartwood. Its grain is usually interlocked, fairly coarse and uneven, with no particular distinguishing features.

Ekki is a heavy, strong wood at around 27 to 32kg (60 to 70lb) per cubic foot when dry. Its density can cause problems when drying. Best practice would indicate that Ekki should be pre-dried for a long period on thin stickers before kilning is attempted. Surface checking and end splitting are the major reasons for degrade, but some distortion will also take place. Fortunately, surface quality is not paramount in most of its eventual end uses. Ekki is rated as very durable – probably the most durable wood out of Africa – it naturally resists any preservative treatment.

Sawing Ekki can be difficult, especially if the material is dry, and severe blunting takes place that can be attributable to the mineral deposits. Regular sharpening or the use of tungsten carbide tips on the saw blades is recommended. Planing can also be troublesome with a lot of chatter if the material being worked is not held firmly. The interlocked, course grain also tends to pick up, so the cutting angle should be reduced to compensate. Ekki is susceptible to splitting, so always pre-drill if nailing and screwing. When boring, it is possible that some charring and burning will take place. It is an

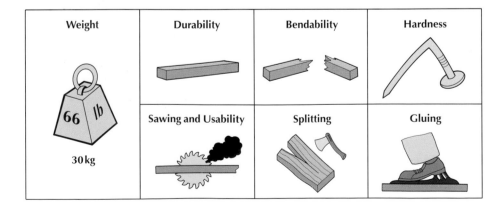

Weight	Durability	Bendability	Hardness
66 lb 30kg	Sawing and Usability	Splitting	Gluing

unsuitable timber to bend being too hard and heavy. Gluing with care can be successful, although results tend to be variable. Staining and polishing are acceptable, but because of the nature of the wood and its eventual end uses these are unlikely to be necessary.

Traditionally, Ekki is cut in large squares and baulk sizes for use in dock work, wharves, bridge-building, sleepers, and so on. In fact, like Greenheart it is very suitable for any outlets where a strong, heavy, hard and durable timber is required. In areas where resistance to wear is vital, such as industrial flooring, it also excels. It has a natural resistance to acids and this leads to use in cider and other fruit presses. So long as surface quality is not important, Ekki has superb qualities that can be put to good use.

Source:
Africa.

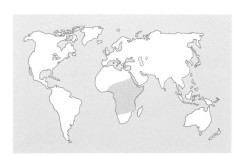

Radial

Tangential

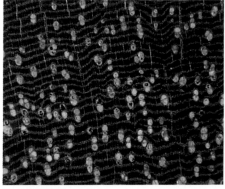

Cross Section

ELM, EUROPEAN AND AMERICAN

Ulmus spp.

Also known as: Nave or Red Elm in Europe and also American, Brown, Cedar, Red, Rock, September Elm, Slippery and Winged in North America.

Lumping all the *Ulmus* species together like this is a little unfair but they have many common features. Since the devastation of Dutch Elm Disease hit traditional sources in Europe little is found except in the far north. Elm's heartwood has a colour that may vary from a dull light brown to a medium brown with a reddish tinge. The sapwood is easily distinguished by its lighter colour. It has a very distinctive grain pattern common to most within the species. Large early wood pores provide the timber with a fairly coarse, cross grain. These are followed by smaller pores at angles throughout the growing season

which produce an attractive wavy pattern. Visible on all surfaces, this feature has a feather-like look and has been called 'partridge breast' figure. Once recognized, this grain pattern is seldom forgotten.

Elm can vary in weight from around 16 to 20kg (35 to 44lb) per cubic foot when dry. The heavier timbers tend to be those called Rock Elm (*Ulmus thomasii*). Although Elm dries fairly rapidly it does have a tendency to distort. This can be alleviated by weighing down the stack or kiln charge with concrete blocks or something similar. Collapse or distortion can be reduced at the end of a kiln run by reconditioning. It is not a strong wood and is somewhat brittle at times. It steam-bends fairly readily, but any timber with knots should be avoided. Elm is not

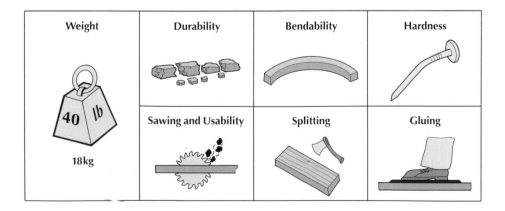

Weight	Durability	Bendability	Hardness
40 lb / 18kg			
	Sawing and Usability	Splitting	Gluing

classified as durable and the heartwood has a moderate resistance to preservative treatment. Some evidence exists that it has an ability to endure when totally submerged in water.

The tendency to collapse when drying plus the irregular grain do cause a certain amount of binding when sawing; it planes reasonably well but care needs to be taken to avoid the grain picking up. Nailing requires little pre-drilling, but do so when screwing. Elm's open grain provides a good surface to bond, although it can lead to problems when staining if a uniform result is required. Good polished surfaces can be achieved with effort.

Traditionally, Elm is used for coffin boards, agricultural implements, pulley blocks, and so on.

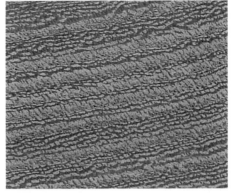

Radial

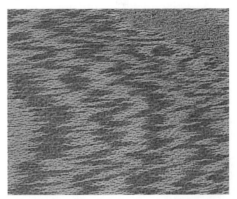

Tangential

Source:

Northern Europe and North America.

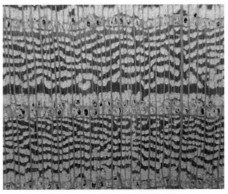

Cross Section

GABOON

Aucoumea klaineana

Also known as: Angouma, Gaboon Mahogany, Gaboon Wood, Mofoumou, N'goumi, Okoume, Ongoumi or Zouga.

Found in a fairly restricted region of Africa, this timber takes its name from the main country of source, Gabon. With a narrow, pale greyish sapwood, the heartwood of Gaboon can sometimes be mistaken for Mahogany, hence the vernacular name. The heartwood can be salmon pink when first cut, but darkens to pinkish brown to light red-brown on exposure. Generally it has a fairly straight grain with some tendency to interlock. The grain does not have particularly distinctive features, although a slight stripe can occur on the quarter-sawn faces. Planed material will often show a lustrous but occasionally woolly surface and cross shakes can also be apparent.

Gaboon is not a heavy timber at around 12kg (26lb) per cubic foot when dry. Nor is it a strong wood and may contain some brittleheart, as can be observed by the appearance of cross shakes on the surface. It dries fairly easily with little tendency to distort, split or check. Gaboon is not durable and resists preservative treatment. However, this is unlikely to be a problem as its typical end uses will not generally expose it to extreme conditions.

The woolly nature of this timber plus the additional problem of silica contained within the pores make it sometimes difficult to saw. Regular sharpening will help to alleviate these problems that also arise when trying to plane the wood. A reduction in cutting angle will help to stop surface tearing and blunting of tools. Nailing properties are good, but as usual it is

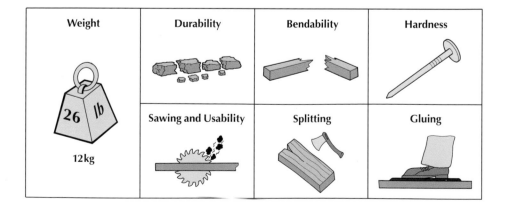

Weight	Durability	Bendability	Hardness
26 lb 12kg			
	Sawing and Usability	Splitting	Gluing

generally helpful to pre-drill when screwing. Gluing is good, staining is acceptable and polishing can produce an excellent finish.

Historically, because Gaboon peels and slices well, it has been found to be a useful timber for plywood, blockboard and other laminated board products manufactured locally. These boards are exported for use in panelling, doors, partitions, and so on. Similar in feel, look and appearance to Mahogany, it has found uses as a substitute for that wood. When it is available, lumber can easily be converted into light internal joinery and mouldings, and it can also be used for furniture and furniture components. It has at times been used for the manufacture of small boxes, especially for cigars.

Radial

Tangential

Source:

West Africa.

Cross Section

GONCALO ALVES

Astronium fraxinifolium

Also known as: Courbaril, Kingwood, Locustwood, Roble Gateado, Tigerwood, Yoke or Zebrawood.

Some of the names used to describe Goncalo Alves are confusing because they are also applied to other species. The tree is found in Central and South America, with most of the small commercial quantities being sourced from Brazil. It grows quite tall, to a maximum of about 46m (150ft). The bole diameters are no more than three or four feet, if left to mature. The lighter coloured sapwood is clearly distinguished from the heartwood, which ranges from a light to dark reddish brown with deeper streaks apparent. Depending upon the plane in which the lumber is cut there may be uniform stripes or irregular ones. It is more uniform when cut on the quarter, but can also have some attractive fiddle-back figure as well. It can occasionally have a mottled appearance with darker spots of colour distributed randomly along the faces of the boards. The grain is most often irregular and interlocked with a medium texture. It is hard and feels cold to the touch.

Goncalo Alves is a dense, hard and heavy wood. It will average anything from 25 to 29kg (55 to 64lb) per cubic foot when dry. As one would expect, it is difficult to dry. Some pre-drying on thin stickers will help to reduce downgrade whilst kilning. Stacks in kilns should be monitored regularly so that surface checking and distortion can be spotted before they become too bad. The wood is very tough and far exceeds Beech in strength values. It is

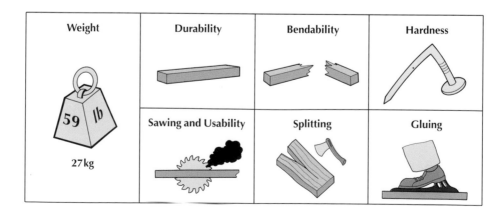

Weight	Durability	Bendability	Hardness
59 lb / 27kg			
	Sawing and Usability	Splitting	Gluing

naturally very durable and will not need preservative treatment.

This is a difficult timber to work. Irregular and interlocked grain may cause saws to bind and planed surfaces to tear out. During planing, the wood will need to be firmly held under pressure to avoid any chattering. It turns well, but frequent sharpening of tools will be required. Pre-drill for all fixing operations. Gluing can be difficult, so try some test pieces first. It is unlikely to be stained, but can, with effort, be polished to a very fine finish.

In Brazil and elsewhere Goncalo Alves is considered to be a highly desirable timber for furniture. Less attractive examples may find their way into structural work. It can be seen used as decorative veneers. Small, fancy turnery and knife handles are favoured applications.

Radial

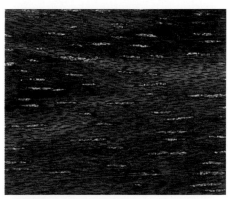

Tangential

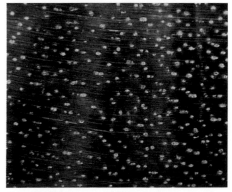

Cross Section

Source:

Central and South America.

GREENHEART

Ocotea rodiaei

Also known as: Black, Brown, Demerara, White and Yellow Greenheart, Groenhart or Sipiroe.

Greenheart is found in the northeast of South America; the main source is Guyana with further smaller amounts available from Surinam. It is very hard, heavy and durable and has therefore been used where these properties can best be exploited. The sapwood is a pale yellow or light green; depending upon the local source it can sometimes be difficult to differentiate from the heartwood. Generally, the heartwood varies from light to dark olive brown or green with, on occasions, some darker streaking. Grain texture is fine and even, with some slight interlocking. Some gum deposits are apparent in the pores and are generally clearly seen on the cross section.

Hard and relatively very heavy, weighing anything from 30 to 36kg (66 to 79lb) on average, per cubic foot when dry, Greenheart is also exceptionally strong. Drying, as for most of these heavy timbers, can be traumatic. Thicker stock should be placed on thin stickers to pre-dry for some time before kilning. Most Greenheart is used in large sections where air drying will be sufficient. Whatever method of drying is used, it takes time and there will be a marked amount of surface checking and end splitting; fortunately little distortion takes place. As one would expect, this timber is classified as very durable and is resistant to preservative treatment.

Working Greenheart can be difficult at times. Stock with interlocked grain tends to distort when cut and can split quite severely. Planing will require sharp tools and in some instances the surface will tend to

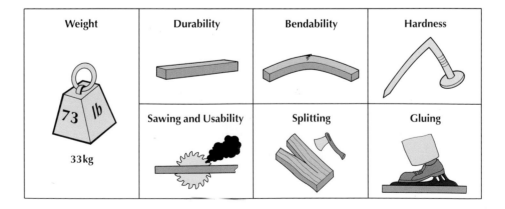

Weight	Durability	Bendability	Hardness
73 lb			
33kg	Sawing and Usability	Splitting	Gluing

flake away. Most of the straight-grained material will produce a good finish with care. Because of its tendency to split rather easily it is advisable to pre-drill for nailing as well as screwing. When boring some charring or burning may occur. Gluing needs care, with some patchy results reported. It is unlikely that there will be a need to stain or polish Greenheart but a good finish can be obtained because of its hardness.

Greenheart is resistant to termites and borers. It is traditionally used for all sorts of wharf and dock works where durability and strength are required. Piers, decks and lock gates are typical. Because it is relatively easy to split, straight-grained stock has historically been used to make fishing rods. Greenheart has some outstanding qualities that are difficult to match.

Radial

Tangential

Source:

South America.

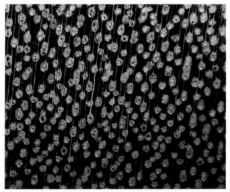

Cross Section

HICKORY, AMERICAN
Caraya spp.

Also known as: Bitternut, Nutmeg, Pecan and Water Hickory, plus Mockernut, Pignut Shagbark and Shellbark Hickory, also Red and White Hickory.

Hickory grows extensively throughout the eastern USA and Canada. Of those listed above, Hickory falls into two broad categories: 'Pecan' Hickory's constitute the first four and 'True' Hickorys the next four. Red and White Hickory are selective names referring to wood from the True Hickory group. The Pecan varieties are not considered to have the same qualities as those of the True Hickorys, although they do have other attributes and are a valuable source of edible nuts. Heavier wood from this group can be utilized for the same end uses as those of the second. Some of the better quality material may find

its way into commercial parcels of timber. The description here relates to the True Hickorys and any of the heavier woods from the first group.

The sapwood of Hickory is wide and a very pale cream or white, and is often known as 'White' Hickory. Many users believe that the sapwood's properties of elasticity and toughness excel over those of the heartwood. A darker, creamy to reddish brown colour, the heartwood can often be found marketed as 'Red' Hickory. The grain is generally straight, but can tend to contain some that is wavy or irregular; it has a coarse texture.

True Hickory varies in dry weight from around 20 to 23kg (44 to 50lb) per cubic foot. Care when drying needs to be exercised. Some pre-drying is advisable to avoid distortion and there can be a tendency for surface checking. Hickory has a reputation for a

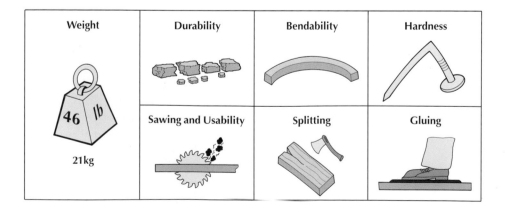

Weight	Durability	Bendability	Hardness
46 lb / 21kg			
	Sawing and Usability	Splitting	Gluing

considerable amount of shrinkage during drying. It is well known for its toughness, hardness, resistance to shock and has strength factors greater than Beech. These qualities do not alter its classification as non-durable with a resistance to treatment.

Depending upon the type of grain structure, Hickory can be sawn and planed relatively easily. The surface of wavy and irregular grained material will have a tendency to pick up. Tools will blunt mildly and it is advisable to pre-drill for jointing functions. It glues, stains and polishes well.

Hickory has found a niche for use in tool handles and similar products where a high resistance to shock is required. In addition to striking tools, it is used extensively for ladder rungs, gymnasium equipment, bentwood furniture components, baseball bats and also decorative veneers and panelling.

Source:

North America.

Radial

Tangential

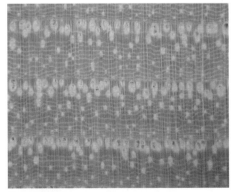

Cross Section

HOLLY, EUROPEAN

Ilex aquifolium

Holly is not often available in large commercial quantities, but it is included here because of its historical use and its fine and even grain qualities. Found in the UK and throughout Europe, it is probably best known as a Christmas decoration. The red berries grow on the female, but not in isolation – a male must be close at hand! The sapwood is a dull, greyish white with little or no demarcation from the heartwood. The poorer material will be found to include some dark, streaky or patchy stains and other defects; typically it will probably have irregular grain. Quality material will have fairly straight grain of a fine and even texture.

A fairly heavy timber at around 23kg (50lb) per cubic foot, it is also quite hard. When drying, it has a tendency to distort rather badly and

as a result is not often kiln-dried unless cut into small sections first. Shrinkage is not a major problem, but this timber can have some dimensional movement after drying when exposed to locations with varied humidity. Holly is not durable, but is unlikely to be used in applications where this might be an issue.

Except for those samples with really bad grain configuration where some burning may occur, Holly can be sawn and planed reasonably well. It will split easily, therefore pre-drill when nailing and screwing. Its gluing properties are good, it takes stain very effectively and uniformly, and can be polished to a fine finish.

Holly has been historically used in situations calling for a close-grained hardwood such as printers' blocks. For centuries, cogs for water and flour mills were made from Oak or Elm

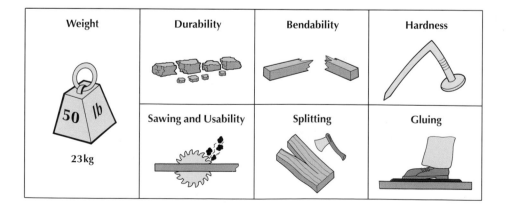

Weight	Durability	Bendability	Hardness
50 lb / 23kg	Sawing and Usability	Splitting	Gluing

with inserted teeth made from Holly and other similar timbers. The word 'treen', meaning 'of trees', refers to such items as drinking vessels, platters or plates and bowls, and Holly was one of the timbers favoured for this use. It was one of the first timbers found in antique furniture to be used for inlay against Oak; it is still used occasionally for this and marquetry. It is additionally used as parts for musical instruments, and sometimes for the butts of snooker and pool cues. It has been stained black and used as a substitute for Ebony, where its fine and evenly textured grain, once polished, can easily be wrongfully identified. It has found favourable use for turning decorative items and is well suited as a material for carving.

Radial

Tangential

Source:

Throughout Europe and similar climates.

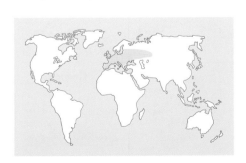

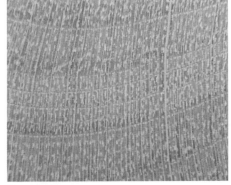

Cross Section

HORNBEAM

Carpinus betulus

The name comes directly from an old German word, 'Hornbaum', meaning hard tree. Although found throughout Europe, most of the commercial supplies are produced in France. Not a large tree, it usually branches low down, which leads to fairly short lengths of lumber. It has an American cousin, *Carpinus americana*, that has little or no commercial significance. There is hardly any difference between sap and heartwood. The timber is a dull white with the possibility of some grey streaks and has a cold feel when first touched. The appearance is fairly featureless, but is generally cross-grained with a close and even texture. When quarter-sawn, it has a large silver grain figure similar to Oak but not as clearly defined.

Fairly heavy at an average of around 21kg (46lb per cubic foot when dry, Hornbeam is slightly stronger than Oak and similar to Beech. It dries well without too much distortion or shrinkage, but is susceptible to movement later if placed in areas of variable humidity. It has good resistance to splitting and has found uses where this attribute can be exploited. Not durable, it will decay quickly where it is continually exposed in damp conditions. Although it will take preservatives well, it is rarely used in applications where this is necessary.

Hornbeam can be difficult to work at times due to the preponderance of cross or irregular grain that may pick up when sawn or planed. Although it has good resistance to splitting it is fairly hard, therefore pre-drilling is recommended when nailing or screwing. It turns well. This timber also has a reputation for resistance to

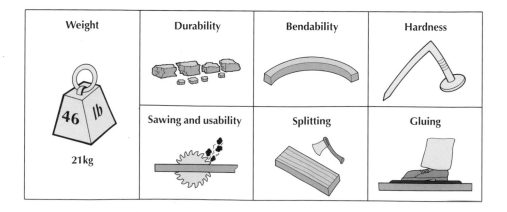

Weight	Durability	Bendability	Hardness
46 lb 21kg	Sawing and usability	Splitting	Gluing

wear. It will steam-bend with relative ease even when small knots are included, although it does have a tendency to discolour. It glues readily, stains with ease and polishes well. Hornbeam generally needs a hard lacquered or varnished finish rather than oil; the latter can look grubby and blotchy.

Historically, Hornbeam has been used in many applications similar to Holly, such as gear teeth, and so on. Because the availability of commercial quantities of Hornbeam today is greatly limited, it is now somewhat of a special-purpose wood. Its resistance to wear and splitting has led to a use for butchers' blocks that has sadly now been superseded by plastic! It is used extensively for piano actions and other musical instrument parts when available. Its attributes have made it popular for hand plane bodies.

Radial

Tangential

Source:

Throughout Europe and similar climates.

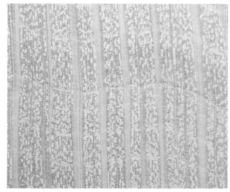

Cross Section

81

IDIGBO

Terminalia ivorensis

Also commercially known as: Emeri or Framire. The tree is sometimes called Black Afara but should not be confused with other species. There are a whole host of vernacular names such as Bona, Cauri, Emil, Frayomile, Mboti, Okpoha, Onhidgo, Ubir and so on.

Idigbo is one of those timbers that has been somewhat overlooked. Found predominantly throughout West Africa, the lumber produced is of a light yellowish, slightly straw colour through to, in some instances, a pinkish yellow. Its texture can be coarse and uneven with some interlocking and wavy grain. It has an unusual growth ring pattern that at times can be mistaken for quarter-sawn Oak. There is little or no difference between the sap and heartwood colour.

It is not a heavy timber at between 16 to 18kg (35 to 40lb) per cubic foot, although there can be some considerable variation from this. Brittleheart, 'cross' or 'thunder' shaking, are apparent in the lighter, core wood. Where these compression failures are not included, Idigbo has a good weight to strength ratio that is slightly below that of Beech. It dries well and fairly rapidly without too much distortion. Once at its required moisture content it is stable with little tendency for further movement. It is considered a durable timber that resists preservative treatment; if this takes place it may leach out afterwards. In exposed or moist conditions when stacked for drying a dark yellow dye can be exuded. This can spoil a planed finish surface if it gets wet, therefore after machining care needs to be taken with storage.

This is a reasonably good material to saw and plane. There can be some

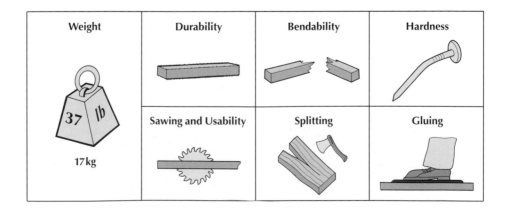

Weight	Durability	Bendability	Hardness
37 lb / 17 kg			
	Sawing and Usability	Splitting	Gluing

surface tearing or pick-up on planed quarter-sawn or interlocked-grain faces. Idigbo does have an inclination towards splitting, so pre-drill when nailing and screwing. If it is likely that the final product will be used in a damp or moist location avoid steel fixings. Acids in the timber will cause a dark stain similar to those found in Oak. It glues well. The open grain may need filling before staining and polishing, but a good surface finish can easily be achieved.

Idigbo has many features that will lend themselves to a variety of uses. The figure-like feature has been used in dark-stained furniture as a replacement for Oak, and may even have been marketed as such! Its stability provides a good interior or exterior joinery product, although not in damp conditions unless sealed.

Radial

Tangential

Source:

West Africa.

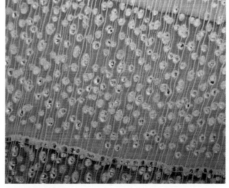

Cross Section

IROKO

Milicia excelsa and regia, (formerly Chlorophora)

Also known as: Abang, Bang, Bangui, Intule, Kamba, Kambala, Lusanga, Mereira, Moreira, Mulundu, Mvule, Odoum, Odum, Rokko, Roko and Tule, to name but a few!

Commonly found throughout Africa from the East to West coasts, *M. regia* has similar features similar to *M. excelsa* but is slightly smaller and only found in Western regions. Iroko is probably one of the best known African hardwoods. It can be bright yellow when freshly cut from a log or large baulk, but it soon darkens to a yellowish brown and eventually becomes a dark golden brown. The sapwood is pale cream and can easily be distinguished from the heartwood. Evidence of light-coloured calcium deposits is often found, and these occasionally become quite large and stone or sheet-like. Interlocked and irregular grain is a feature of this wood and it has rather a coarse, open appearance. It can, at times, have an oily feel to it until thoroughly dry.

A medium-weight timber at around 18kg (40lb) per cubic foot when dry, it is similar in strength to Beech. Iroko dries fairly well without too much distortion or splitting, although bad consignments do occur in which both of these can be major problems; wider boards will benefit from having their ends cleated to help avoid splitting. Some case-hardening can also take place and it is advisable to condition kiln loads prior to discharge to avoid problems when sawing. Dry stickers should also be used to alleviate any likelihood of finding stains or 'shadow' marks on boards when planed. The heartwood is classified as very durable and resistant to preservative treatment.

Iroko will normally saw and plane

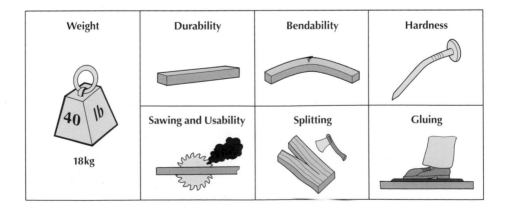

Weight	Durability	Bendability	Hardness
40 *lb* / 18kg	Sawing and Usability	Splitting	Gluing

reasonably well, although if calcium is present expect to resharpen often. The presence of interlocking grain combined with any case-hardening could lead to burning and spectacular distortion off the saw! Material with poor grain quality will also pick or tear out when planed. It is always advisable to pre-drill this wood when nailing or screwing. Although there is sometimes an oily feel to the surface, Iroko does glue well. Some surface filling may be necessary before staining and polishing, but a good finish is normally obtainable.

This strong, durable and attractive wood has found many end uses, typically in high-class joinery, garden and house furniture, doors, work tops and other similar applications. It is a cheap substitute for Teak and because of its extensive use in Eire is flippantly known as 'Irish Teak'!

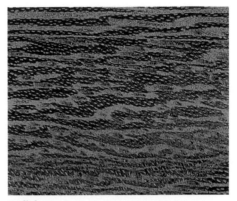

Radial

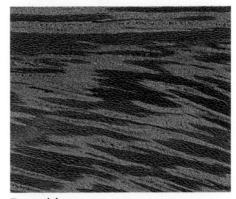

Tangential

Source:

Africa.

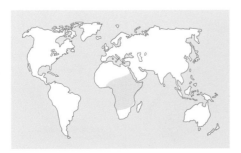

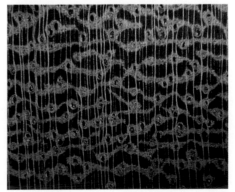

Cross Section

85

JARRAH

Eucalyptus marginata

Sometimes also known as: Curly Jarrah.

One of Australia's well-known Eucalypts, Jarrah is a popular commercial timber where strength, durability and resistance to wear are required. However, abundant commercial quantities are usually confined to South-west Australia. Freshly sawn timber can have a pinkish hue darkening to a deep, rich mahogany-type red. Some darker flecks can be seen on most faces, adding to its appeal. The sapwood is fairly narrow and paler in colour. The grain tends to be even but slightly coarse; it is straight in most instances but an attractive, short, wavy or curly grain is often found. Gum can be present in the form of pockets or veins.

Jarrah is a fairly heavy wood at around 23kg (50lb on average per cubic foot when dry. It is also strong, being comparable to Beech, and has good resistance to wear but does tend to splinter. As it is with most heavy timbers, drying can be difficult, so a period of pre-drying, especially of the thicker material, is advisable to help this along. Shrinkage factors from green to dry can be high, with some distortion taking place. There is some tendency towards surface checking and end splitting. This is a very durable hardwood that resists treatment with preservatives. Its heartwood is also credited with being resistant to termite attack!

This can be a difficult timber to saw and plane, especially the curly or interlocked-grain material. Some surface tear-out is to be expected, which can be eased by reducing the planing cutting angle. To avoid too many bent nails it is best to pre-drill

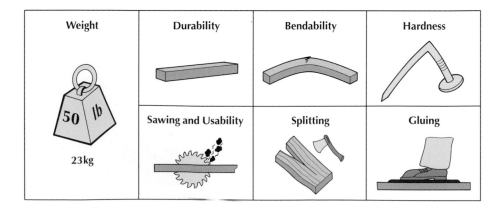

Weight	Durability	Bendability	Hardness
50 lb 23kg	Sawing and Usability	Splitting	Gluing

Jarrah; naturally, the same applies when screwing. It has good gluing properties. Not often stained, it will produce a fine polished surface.

Jarrah has exceedingly good mechanical properties, and many of its end uses have reflected this. It is often used in similar structural situations to Greenheart, for decking, dock and pier work. It has also been used extensively for railway sleepers. It is a good flooring material, although it is probably best used for domestic purposes as it tends to splinter in locations of heavy wear. It can easily be used for internal and external joinery and cabinet work. Logs containing the curly grain are sometimes cut and used for veneers. In recent times, Jarrah has found a particular use in garden furniture, trellis work and similar end products.

Radial

Tangential

Source:

Western Australia.

Cross Section

JELUTONG

Dyera costulata and lowii

Also known as: Andjarutung, Djelutung, Jelutong Bukit, Jelutong Paya, Letung or Tinpeddaeng.

Jelutong is found throughout Malaysia and Indonesia. It is a lightweight timber with an off-white colour when freshly cut. After exposure, this darkens slightly, with a yellowish straw-like tinge; there is no differentiation between the sap and heartwood. Jelutong is a plain timber and there are no discernible grain features. It is usually straight, with a fine and even slightly lustrous texture. Latex traces are generally present and appear as slit-like radial passages that are spaced at intervals along the grain. This is a useful resource for which the tree is tapped and the latex collected for use in the manufacture of chewing gum!

On average, when dry, Jelutong will weigh about 13kg (29lb) per cubic foot. It is not a strong wood and is easily dented; it can be quite brittle at times. Classified as non-durable, this wood can easily be treated with preservative if so desired. It is susceptible to fungal attack when wet therefore needs to be stacked on stickers fairly quickly after conversion. Dipping or spraying with an anti-stain fungicide after conversion helps to prevent this, but avoid the splinters! It dries quickly and well with little distortion and only some occasional evidence of surface checking. Careful choice of sticker will help to avoid any further staining during this process. Once dry, if used in favourable conditions this timber is fairly stable and will keep its shape well.

As a fairly soft timber it is easily sawn, planed and moulded. However, care does need to be taken to ensure that all tools are sharp to provide a

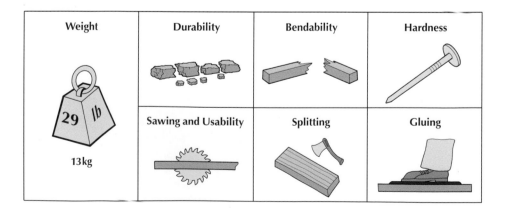

Weight	Durability	Bendability	Hardness
29 lb 13kg			
	Sawing and Usability	Splitting	Gluing

good surface finish, especially if latex is present. It turns easily and carves well, properties which help to determine some of its eventual uses and applications. It has a natural resistance to splitting and can be nailed with ease. It glues well and accepts stains readily with an ability to produce a good finished surface.

As a lightweight, easily worked and stable timber, Jelutong has found a use as a pattern-making material. These attributes also lend themselves to use in craftwork as a carving medium. The stability feature is also good for drawing boards where the wood is best laminated to produce the wider widths. As a light material, it is often found as a core stock in flush doors, plywood and panelling.

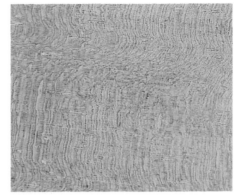
Radial

Tangential

Source:

South-east Asia.

Cross Section

KAPUR

Dryobalanops spp.

Also known as: Borneo Camphorwood, Brunei Teak, Enteng, Kapoer, Kapor, Kapur Paji, Malampait, Santjulit, Sintok or Tulai.

Found throughout Malaysia and Indonesia, most of the commercial supplies of Kapur come from Sabah or Sarawak. As the alternative name Borneo Camphorwood suggests, there is a distinct smell of camphor about the wood when freshly cut, which goes off after drying but can still be noticeable when the wood gets wet. The sapwood is pale yellow or light brown and is easily distinguishable from the heartwood, which is a light to darker reddish brown. The grain is usually straight, with some interlocking of a fairly coarse but even texture. Fine resin ducts are present and visible to the naked eye. The resin does not exude onto the surface as in Keruing, but there can sometimes be a slight oily feel to the wood. Another feature is a tendency to leach out a dark stain when the timber is stacked in wet conditions. This staining can be exacerbated when in contact with ferrous metal. On occasions, some parcels of Kapur can be badly affected by pinhole borer damage.

Kapur is a medium to heavyweight timber and can vary depending upon source. Generally, when dry, it will be between 20 to 23kg (44 to 50lb) per cubic foot, averaging 21kg (46lb). It dries slowly, with some tendency to cup. Any shakes or splits already present in the wood prior to kiln drying will probably get worse. Thick material takes some considerable time to dry, with every likelihood of surface degradation due to checking. It is not commonly used in thickness over 5cm (2in) to avoid this problem. Kapur is

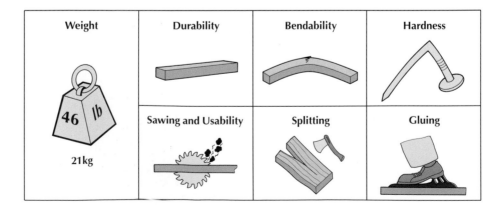

Weight	Durability	Bendability	Hardness
46 lb 21kg			
	Sawing and Usability	Splitting	Gluing

rated as slightly stronger than Beech. It is very durable and resists treatment with preservatives.

Sawing Kapur can sometimes be difficult; severe blunting can occur, especially when sawing wood that is heavier with more interlocked grain. Spelshing on the undersides of sawn surfaces is also apparent. Planing tools need to be sharp in order to maintain a smooth surface and avoid the grain tearing or picking up. Easily split, Kapur needs pre-drilling for all fixing operations. This is a utility timber and will not often need to be stained or polished. Gluing should not be problematical.

Kapur finds employment most frequently in constructional-type work, but also has some uses in domestic and light industrial flooring, trailer beds and external joinery.

Radial

Tangential

Source:

South-east Asia.

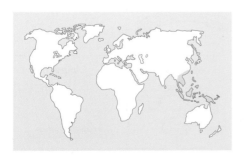

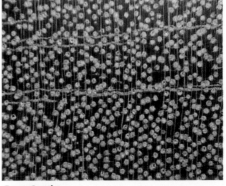

Cross Section

91

KARRI

Eucalyptus diversicolor

Similar to Jarrah, this timber is an indigenous species found mainly in South-western Australia. It is a hard and heavy wood with many of the same characteristics. This tree is one of Australia's largest, growing anything from over 46m (150ft) to considerably in excess of 60m (200ft) high. The sapwood is pinkish, while the heartwood is a reddish brown that generally tends to be slightly lighter than Jarrah in colour. It is often difficult to tell the difference between these two woods, although adopting the 'ash' test will help. Burn a sliver about the size of a match and inspect the residual ash; if it is white this indicates it is Karri; if of a black, charcoal-like appearance it will be Jarrah. Karri has fairly straight or slightly interlocked grain that can produce a striped effect; it is coarse but evenly textured.

Heavier than Jarrah this timber, when dry, will be between 25 to 27kg (55 to 60lb) per cubic foot, 26kg (57lb) on average. It is tough as well as heavy, with strength values well above those of Beech. Drying can be very difficult with large amounts of deep surface checks appearing, especially in the thicker stock sizes. Pre-drying is advisable on thin stickers to help alleviate this problem. In addition to checking, it will also tend to end split and distort. There is a fair amount of shrinkage while drying takes place. Karri's weight and toughness do not translate into durability – in locations of continual exposure to damp or variable conditions and in contact with soil it does not remain sound for extended periods.

Karri's weight and inclination towards interlocked grain tend to make sawing difficult. Sharp tools are

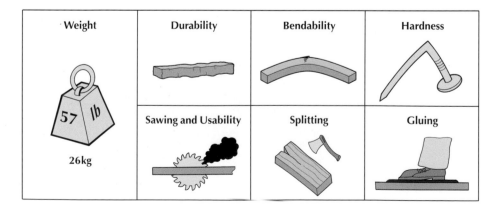

Weight	Durability	Bendability	Hardness
57 lb / 26kg			
	Sawing and Usability	Splitting	Gluing

a necessity and you can expect to resharpen more often than with other timbers. Reducing the cutting angle when planing will help to negate any surface degradation due to tearing or picking up. Nailing and screwing are difficult and pre-drilling is necessary. Staining and polishing can both be achieved relatively easily and Karri glues well.

This timber has found many structural and construction uses where strength is required, for example bridge-building, beams, joists and rafters. It can also be used in many other applications such as in industrial flooring, wagon beds and structural plywood. Locally, it is used for internal house finishing and furniture. It should not be used in positions where it will be exposed to moisture for extended periods.

Radial

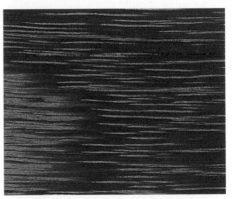

Tangential

Source:

Western Australia.

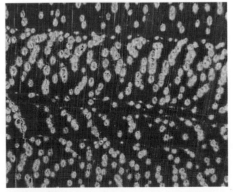

Cross Section

KERUING

Dipterocarpus spp.

Also known as: Apitong, Bajac, Dau, Eng, Engurgun, Gurjun, Hieng, In, Keruing Belinbing, Minyak Keruing or Yang.

Keruing and other vernacular names are used for a grouping of many species of *Dipterocarpus* ranging throughout India, Myanmar (formerly Burma), Thailand and the rest of South-east Asia. In recent years, 'Dau' from Vietnam has also become available. Most commercial quantities of Keruing come from Malaysia and Indonesia. Here, the number of species is greater, giving more variety to the wood than from other sources. The sapwood is grey in colour and distinct, often containing evidence of wood borer attack. The heartwood varies from a pinkish brown through to a dark brown. The grain is coarse, open, interlocked but fairly even; and typically covered with a resinous

exudation that is sticky to the touch. Some batches with fewer numbers of species included, such as Yang or Gurjun, may not be so obviously sticky.

Weight is variable, but an average dry weight of around 21kg (46lb) per cubic foot is usual. Uniform drying can be slow and difficult to achieve in a kiln. It will cup badly and distort generally; those species with resins present will end up with hard globules of the stuff over all surfaces! Even when air-dried, Keruing is not particularly stable; it responds by expanding and contracting with the seasons. Once dry, it is considered to be a strong timber, somewhat stronger than Beech. Because of the mix of species, Keruing as a group is classified as moderately durable.

Sawing Keruing can be difficult; the saw's teeth tend to clog up with

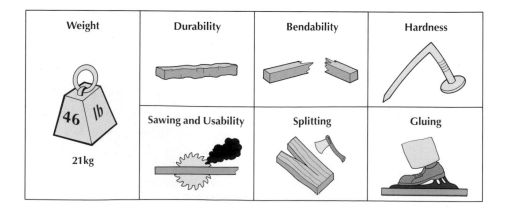

Weight	Durability	Bendability	Hardness
46 *lb* 21kg	Sawing and Usability	Splitting	Gluing

resinous deposits and need cleaning regularly. A slightly fibrous finish is usual when planed, and once more the resin can be a problem. Keruing is easy to split, so pre-drill everything before nailing or screwing. Gluing can be variable and care is needed in critical situations. Staining and polishing Keruing is unusual and does not produce good results.

A medium to heavyweight timber, Keruing has found many uses in constructional applications. Not one of the best timbers for joinery, it has been used for threshold sills in door frames. Any location that is exposed to sunlight should be avoided, as the resin will continue to come out even through painted or finished surfaces. A lot of thinner, smaller sections are used in furniture for stuff over work where appearance is not critical.

Radial

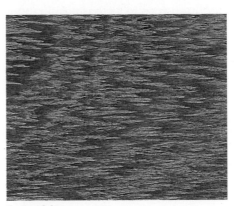

Tangential

Source:

India, Myanmar and South-east Asia.

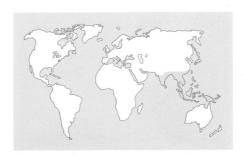

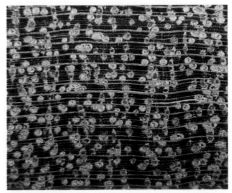

Cross Section

95

LIGNUM VITAE

Guaiacum spp.

Also known as: Bois de Gaiac, Guaiacum Wood, Lignum Sanctum Pokhout or Porkholz.

Lignum Vitae can lay claim to being one of the heaviest and hardest commercially available woods on the market. The main species is *G. officinale* found throughout central and northern parts of South America; this is known as 'true' Lignum Vitae. Other species such as *G. sanctum* match its properties closely but are often referred to as 'bastard' Lignum Vitae. The tree, whose name means 'wood of life', was prized for medicinal properties believed to be contained in its resin. A small tree of no more than about 9m (30ft) high, the trunks produce billets of wood from 7.5 to 30cm (3 to 12in) in diameter; larger sizes are rare. The 'true' wood has a narrow, very distinct band of sapwood while the 'bastard'

wood is generally much wider and can represent the greater volume of the log. The heartwood is typically a dark green or nearly black colour, sometimes streaked. It contains resins that feel oily to the touch and provide it with self-lubricating properties. When freshly sawn, warm or rubbed it has a pleasant smell. The grain has a fine and even texture which is interlocked.

Lignum Vitae averages around 37kg (81lb) per cubic foot. It is extremely tough and resistant to wear, but can be brittle under impact. It is a difficult wood to split radially; tangentially it is relatively easy. Care is needed when drying – if exposed to high temperatures some surface checking and, possibly, ring-shakes may develop at the ends. Pre-drying on thin stickers is recommended before kilning. Lignum Vitae has a

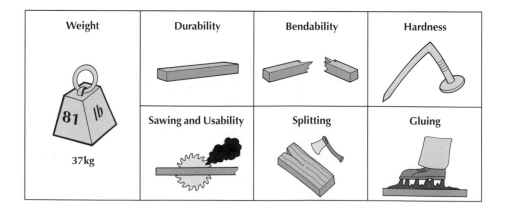

Weight	Durability	Bendability	Hardness
81 lb / 37kg			
	Sawing and Usability	Splitting	Gluing

natural resistance to most fungal and wood-boring attacks and is therefore classified as very durable.

Due to the refractory nature of the wood it is difficult to saw and plane; a reduction of cutting angles will help. It does turn exceedingly well, although tool sharpness needs to be maintained. If nailing or screwing Lignum Vitae it is a good idea to pre-drill to avoid too many bent or broken ones! The resinous content of the wood makes it difficult to glue. It polishes to a fine finish.

Lignum Vitae's self-lubricating properties have led to a use for bearings in ships, where it can outlast metal. It is used in turnery for fancy items and for sports goods, especially bowls for flat and crown green bowling.

Radial

Tangential

Source:

Central and northern South America.

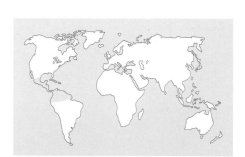

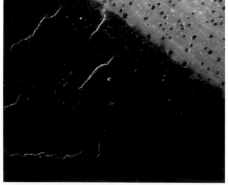

Cross Section

MAHOGANY, AFRICAN

Khaya spp.

This Mahogany is often prefaced with the name of its source country, for example Nigeria, or port of shipment, for example Lagos Wood. It is also called: Acajou Blanc or Acajou D' Afrique, Khaya, Krala, Ogwano or Ngollon.

Found mainly in the tropical rainforests of West Africa, the tree *K. Senegalensis*, a darker heavier wood often marketed as 'Heavy' Mahogany, also grows in the dryer zones. The description in this book focuses on the tropical source from which most of the commercially available material comes.

A large tree, African Mahogany can reach over 60m (200ft) with clear bole diameters of 1.5 or 1.8m (5 or 6ft). It has a pleasant pink colour when freshly cut which soon turns to a typically darker reddish brown. The sapwood is not always easily discernible from the heartwood. The grain is generally interlocked, producing a striped effect on quarter-sawn faces that is similar to Sapele; it has a fairly coarse texture. On close inspection, some evidence of brittleheart may be noticed. It is often difficult to see until the timber is planed, when it is displayed by fine cross-shaking usually along one edge of the board.

A fairly lightweight timber, parcels of African Mahogany can vary due to the mix of species or origin. On average when dry it will be around 15kg (33lb) per cubic foot. It will dry fairly rapidly without too much distortion unless a preponderance of tension wood is present; shrinkage is minimal. It is not a strong timber, especially those examples with brittleheart, but it does have a good weight to strength ratio. However, its

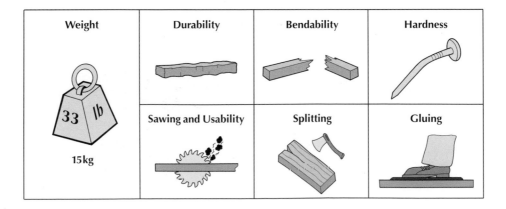

Weight	Durability	Bendability	Hardness
33 lb 15kg			
	Sawing and Usability	Splitting	Gluing

surface is easily dented with a finger nail. Classified as moderately durable, it will resist any treatment with preservatives, although fortunately it is not often used in applications where this is needed.

Machining African Mahogany can be difficult at times where tension wood is present. Woolly planed surfaces often occur in quarter-sawn boards but generally it produces a smooth, fine finish. Nailing and screwing are both satisfactory and it glues well. Often stained, African Mahogany accepts it uniformly. When polished, it produces a good surface finish.

African Mahogany was first used as a substitute for Cuban or Honduras Mahogany and is closely related. Now it is rightfully used on its own merits for furniture, fine cabinet work and decorative veneers.

Source:

West Africa.

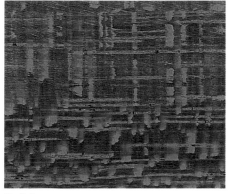

Radial

Tangential

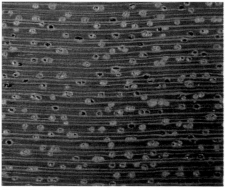

Cross Section

MAKORE
Tieghemella heckelii

Also known as: Abaku, Agamokwe, Baku or Cherry Mahogany.

Similar in some ways to close-grained Mahogany, this African redwood is found predominantly in the western regions. Occasionally incorrectly called Cherry Mahogany or African Cherry, these names are misleading and not often used today. The heartwood varies in colour from a pinkish purple-red through to a deeper blood red-brown, with the possibility of occasional darker streaks. In some examples, chequered and wavy figures are also apparent. In contrast, the sapwood is well defined as a pale yellow or cream colour 5 or 7.5cm (2 or 3in) wide. When planed, the surface is much finer than Mahogany, with a finish that is clearly lustrous.

This is quite a hard and heavy redwood at around anything from 18 to 21kg (40 to 46lb) per cubic foot, averaging 19kg (42lb). It is a dense, tough wood that is fairly stiff, with strength values slightly less than those of Beech. It dries with some slight distortion and at a reasonably slow rate. If knots are present, it occasionally splits around them. Some twisting may occur, therefore it is good practice to weight down the top of a kiln charge to help to avoid this. Shrinkage is not a major issue and Makore is fairly stable in favourable conditions. Its heartwood is classified as very durable and resists all preservative treatments with vigour!

The silica content in the structure of the wood blunts tools quickly, especially if cutting dry wood, and resharpening can become a bit of a chore. However, with a sharp edge, good surfaces can be achieved. Pre-drill for all fixings; the wood has a

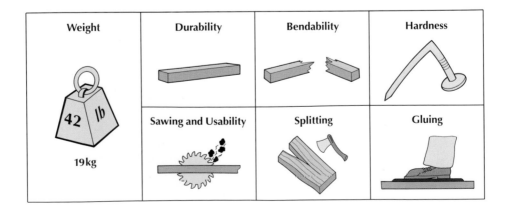

Weight	Durability	Bendability	Hardness
42 lb / 19kg			
	Sawing and Usability	Splitting	Gluing

tendency to char when bored. It glues well. Stains are accepted readily and a good surface polish can be achieved. Makore is often peeled and sliced for veneers, where the higher moisture content makes it an easier material to cut.

Most commonly used for furniture and cabinet work, Makore is also ideal for all internal and external joinery purposes. It can stain in damp conditions when in contact with ferrous metals; care therefore needs to be taken with fixings. As a durable and relatively strong timber, Makore is used in keels, decking and framing for boat-building; as a plywood, its durability lends itself to use in marine and body-building applications. It is also suitable for use as a flooring material, preferably in domestic rather than industrial use.

Radial

Tangential

Source:
West Africa.

Cross Section

103

MANSONIA

Mansonia altissima

Also known as: Aprono, Bete or Ofun.

Mansonia is another timber from the countries of Western Africa. The tree grows in the tropical rainforests to about 30m (100ft) high. It does not have a large bole; 60cm (2ft) or more is considered exceptional. It has occasionally in the past been offered commercially as African Black Walnut. This a confusing title because there is already an 'African Walnut', *Lovoa trichilioides*, which is dealt with later. In addition to this, Mansonia does not belong to the Walnut family. It is possible that the grain characteristics once led to this former inappropriate name. The sapwood is clearly defined and is creamy white in colour. In contrast, the heartwood varies from a medium yellowish brown through mid-brown to almost black. Initially, there may be some purple streaking or tinges that will tone down and fade after exposure. The grain is normally straight with a fairly fine and uniform texture.

Mansonia is a medium-weight timber at about 17kg (37lb) on average per cubic foot when dry. It is fairly hard and is considered equal to Beech in strength values. It dries fairly rapidly with only minimal shrinkage, but does have a slight tendency to warp. There is a likelihood that existing splits will extend and some may also appear around any knots included in the lumber. Mansonia is classified as very durable with extreme resistance to preservative treatment.

This timber has good sawing and planing qualities, cutting cleanly without much lifting or tearing out of the surface grain. Only moderate blunting of machine tools takes place. Working with hand tools is also fairly easy and is pleasant on all three faces.

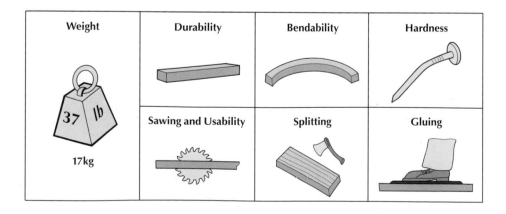

Weight	Durability	Bendability	Hardness
37 lb / 17kg			
	Sawing and Usability	Splitting	Gluing

It is classified as a good material to steam-bend, but avoid knots of any description when doing this. It takes nails reasonably well, but pre-drill when screwing. No problems are usually experienced when gluing. If required, it will take stains readily and a good polished finish is easily achieved.

The darker coloured wood of Mansonia has sometimes been used in applications similar to Walnut. In its own right, it is a very suitable timber for furniture, cabinet work, panelling and interior decorations. Piano cases have been made from it where economy plays a key role. It can be used with confidence for most internal and external joinery, with little movement in favourable conditions. In local countries of origin it is used in plywood production.

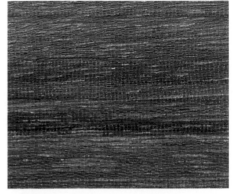
Radial

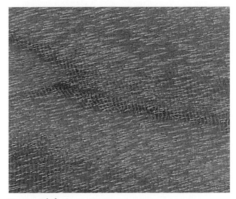
Tangential

Source:

West Africa.

Cross Section

MAPLE, ROCK

Acer saccharum and *nigrum*

Also known as: Black, Hard, Sugar or White Maple or Black Sugar Maple.

This is the heavier grouping of North American Maples and is most commonly marketed as Rock or Hard Maple. It is from these trees that the valuable Maple sugar and syrup are extracted. The sapwood is generally a light creamy white with little or no differentiation from the heartwood. Light or pale cream, the heartwood can contain a reddish or brownish tinge. The wood of *A. saccharum* is often attacked by an insect that can have a great effect on the wood. The results of this invasion become the finely figured 'Bird's Eye'. The grain can be wavy or curly but generally tends to be straight. It has a close, fine and even texture with slightly darker brown lines showing the growth rings that can be clearly seen on the tangential surface. In addition to Bird's Eye material, Maple is sought after for attractive curly or fiddle-back veneers.

This group weighs on average, when dry, around 21kg (46lb) per cubic foot. It dries fairly slowly without too much distortion or downgrading. It is well known that shrinkage can be significant during the drying process. The wood is quite heavy, hard and is resistant to shock; it is rated comparable to Beech in strength. This is not a durable wood and is seldom used in external applications; it also resists preservative treatment.

Sawing Rock Maple is generally reasonable but some blunting will take place. When cross-cutting, it is usual for the end grain to show signs of burning. The hardness of this wood causes some chattering when planing takes place; be sure to machine under pressures and with firm hand control.

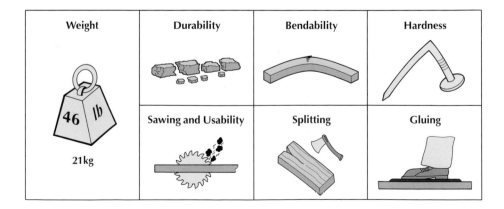

Weight	Durability	Bendability	Hardness
46 lb / 21kg			
	Sawing and Usability	Splitting	Gluing

It is reasonably receptive to nails, although it is probably a good idea to pre-drill for this and screwing. It glues well. Not usually stained, it will accept them readily and can be polished to an excellent finish.

The exceptionally attractive nature of some of the figure in this wood has lead to extensive use for veneers in high-class furniture, cabinet-making and panelling; solid wood furniture is also very popular. Its strength, resistance to wear and abrasion properties are exploited for flooring in sporting facilities such as squash courts, skating rinks, sports and school halls. It turns well and is used for handles, rollers in the textile industry and other similar applications.

Radial

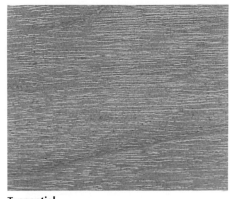

Tangential

Source:
North America.

Cross Section

MAPLE, SOFT

Acer saccharinum, rubrum and *negundo*

Also known as: Red, River, Scarlet, Silver, Swamp, Water or White Maple and Boxelder.

There are a confusing number of names applied to this group of Maples; in addition, 'White Maple' is also a name by which Rock Maple is known. Generally, they are all lumped together and marketed as Soft Maple. The wood is very similar to that of Rock Maple but is significantly softer. It is often confused with Sycamore, (*A. pseudoplatanus*). Field Maple (*A. campestre*), and Norway Maple (*A. platonoides*), which can all be found growing from West Asia, through Europe to the UK.

The sapwood is a creamy white blending with little or no demarcation between it and the heartwood. The growth rings are lighter in colour and therefore are not particularly distinct; the grain is usually straight and not renowned for its figure. Soft Maple does frequently have a display of pith flecks on some surfaces that, depending upon your view, can be considered a feature.

This is not a hard or heavy wood at around 16kg (35lb) per cubic foot when dry. It is not a strong timber, with values rated well below that of Beech. It dries slowly, in a similar fashion to Rock Maple, without too much degrade. Although still readily apparent, there is not such a significant amount of shrinkage. Classified as non-durable, this timber is not particularly suited for external use. If it is to be used in exposed or inclement conditions it is slightly more receptive to preservative treatment than the Rock Maples.

Because it is softer than Rock Maple, Soft Maple's working properties are slightly better. Saw

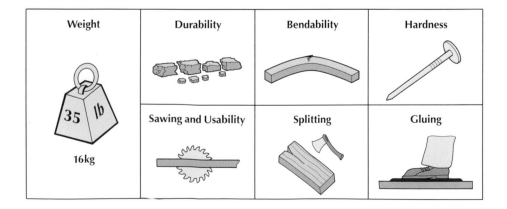

Weight	Durability	Bendability	Hardness
35 lb 16kg			
	Sawing and Usability	Splitting	Gluing

blades and planer irons need to be kept sharp to maintain a good and satisfactory surface finish. There is less likelihood that chattering will take place. Steam-bending Soft Maple is normally successful but knots and curly or uneven grain should be avoided. Although a softer Maple it is best to pre-drill before screwing to avoid splits, especially near the end of a work piece.

Soft Maple does not possess the same resistance to wear and abrasion as Rock Maple and is therefore not often used for flooring. Lumber and veneers can be substituted as a cheaper alternative in furniture manufacture. Its ability to turn well results in extensive use for handles of kitchen and utility utensils. Not recommended for use in external joinery unless treated with preservative.

Source:

North America.

Radial

Tangential

Cross Section

MELUNAK

Pentace spp.

Also known as: Burmese Mahogany, Kashit from Myanmar (*P. burmanica*), and Thitka.

This tree can be found in Myanmar and throughout South-east Asia. Thitka is most common from Myanmar and Melunak from Malaysia, of which *P. tripera* is the predominant species. Large commercial quantities are not available from either of the main sources. The timber from Myanmar may have a slightly closer grain structure than from other countries. In general, the sapwood is visible but not always clearly defined from the heartwood. It is lighter initially than the latter, which tends to be pink-brown through to a reddish brown that darkens on exposure. The grain is interlocked and displays a stripe or roe-like figure on quarter sawn faces; the texture is reasonably fine and even but can be quite open at times. A white deposit is occasionally visible within the pore structure.

Melunak's dry weight is around 18 to 21kg (40 to 46lb) per cubic foot, averaging 19kg (42lb). It dries reasonably well but slowly. There can be some tendency for surface checking to appear on thicker material and some warping will be apparent. With care, the degrade should not be significant. Melunak's strength is rated slightly less than Beech. The timber is classed at the bottom end of moderately durable; treatment with preservative can be difficult but should be applied for outside applications.

The interlocking and slightly irregular grain coupled with any white deposits make this a fairly difficult timber to work. Cross-cutting may produce burning and surface tearing;

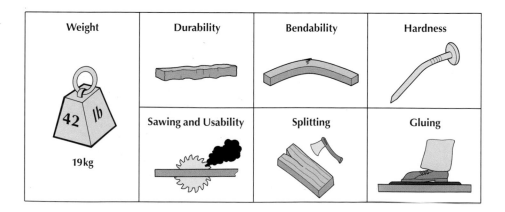

Weight	Durability	Bendability	Hardness
42 lb 19kg	Sawing and Usability	Splitting	Gluing

pick-up during planing is common. To help improve this latter situation, reduce the cutting angle of planer knives. Melunak will turn reasonably well with sharp tools. Although there is some resistance to splitting, it is advisable to pre-drill when nailing and especially when screwing. Melunak glues in a satisfactory manner. Open-grained material will require a certain amount of filling before final surface finishing takes place. When well prepared it takes stains evenly and polishes very well.

Local uses include mathematical instruments such as squares, measures, straight edges and boxes. It is sought after as a timber for high-class furniture, especially in Myanmar. It has been used in boat-building, shopfitting and interior and exterior joinery. It is also a good material from which to make walking sticks!

Source:

Myanmar and South-east Asia.

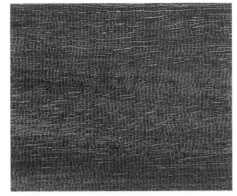

Radial

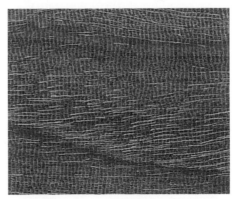

Tangential

Cross Section

111

MENGKULANG

Heritiera littaralis and *simplicifolia,* (formerly *Tarrietia spp.*)

Also known as: Chumprag, Chumprak, Dungan, Kembang, Lumbayan or Palapi.

Found throughout South-east Asia, timber from this tree is not as widely available as it was a number of years ago. It is quite a large tree, producing lumber in long, straight lengths free from knots. Its light brown or slightly yellowish sapwood is clearly distinguishable from a darker reddish brown heartwood. Mengkulang has the appearance of open, coarse but fairly even grain although some interlocking is present, creating a stripe-like effect. Dark streaks are apparent on some faces. There is a distinct short, red fleck found on quarter sawn faces and speckles on tangential faces; both of these are caused by the exposure of the ray structure. When first cut, the wood may have a slightly unpleasant smell but this soon goes off. It can feel oily or slightly greasy to the touch. The overall appearance is very similar to that of Niangon from West Africa.

Mengkulang is a medium-weight timber at between 19 to 21kg (42 to 46lb) per cubic foot when dry, averaging 20kg (44lb); there can be some variations from this base depending upon source. Overall, its strength values are comparable to Beech. Some noticeable shrinkage takes place during drying that may lead to distortion if the timber is not stacked correctly. It dries fairly quickly, but can warp and twist with some occasional surface checking; in some instances end splitting is a problem. The heartwood is not rated as durable, so care should be exercised if used for external applications; there is a resistance to preservative treatments.

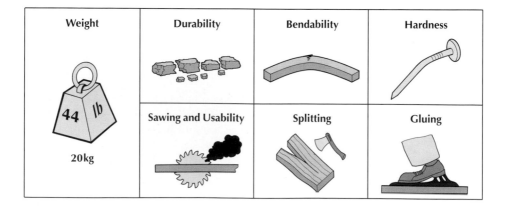

Severe blunting of tools often takes place when working Mengkulang; this is caused by silica contained within the ray structure. Planer blades tipped with tungsten carbide should be considered if large quantities are to be machined. A reduction of the cutting angle helps to improve surface finish, especially on quarter-sawn material. With a tendency to split, it is wise to pre-drill when nailing and screwing. The open grain structure will accept glue well but care may need to be exercised. This is a utility timber that is unlikely to be stained or polished. If it is, the surface will need to be filled before final finishing takes place.

A fairly attractive wood that has been used for shopfitting, interior joinery, some furniture and trims. It has found some use for decking, planking and coachwork.

Radial

Tangential

Source:

South-east Asia.

Cross Section

MERBAU

Intsia spp. (especially I. bijuga).

Also known as: Borneo Teak, Ipi, Kajubesi, Kwila, Merbau Darat or Mirabow.

Often confused and mistaken for African Afzelia, Merbau was for some time included in the same genus. It is found throughout South-east Asia. When available, commercial quantities are mostly sourced from Malaysia and Indonesia. Quite a large tree, it is not always straight, nor is it particularly plentiful. A wide, creamy white sapwood is easily discernible from the heartwood, which starts out as an orange-brown or medium brown colour that darkens to a uniform brown. Some dark and yellow sulphur-like deposits are often seen on planed surfaces in the open pore structure. The yellow deposits can be extracted for dye. There is a tendency for these and the darker deposits to leach out if the timber is stacked in open, wet conditions. The grain is coarse and open with some interlocking; it may appear wavy or stripy on some faces.

A fairly hard and heavy wood that averages over 23kg (50lb) per cubic foot when dry, this is heavier than Afzelia, with which it is often compared. Its strength is rated slightly weaker than that of Beech and it is quite brittle. It dries reasonably well without too much shrinkage and distortion; wood from bent tree boles may tend to move more. There can be a noticeable amount of end splitting and surface checking. Merbau is classified as a durable timber that is highly resistant to treatment with preservatives.

Merbau can be a difficult timber to work and machine. Blunting while sawing is moderate, but gum does collect on the blades. The wavy and interlocked grain can cause problems

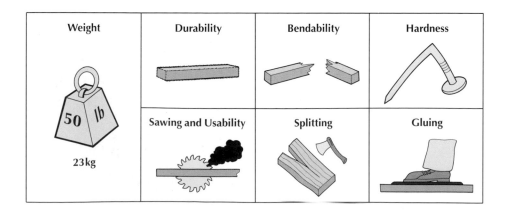

Weight	Durability	Bendability	Hardness
50 lb 23kg	Sawing and Usability	Splitting	Gluing

when planing due to tearing and picking up of the grain. Reducing the cutting angle will help to alleviate this, as will keeping all tools sharp. It is not a good timber to nail with plenty of inclination towards splitting, pre-drill for this and for screwing. The open grain structure helps when gluing but will need filling before any surface finishing takes place. If necessary, Merbau can be stained and will polish up to a fine surface gloss.

This is a useful material for flooring in both industrial and domestic applications. If used for external joinery it will need to be well sealed to avoid any stain leaching out. Some of the more attractive timber can be made into furniture, panelling and interior trims.

Radial

Tangential

Source:

South-east Asia.

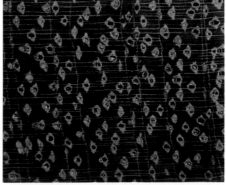

Cross Section

MUNINGA

Pterocarpus angolensis

Also known as: Ambila, Bloodwood, Kajat, Kajatenhout, Kejaat, Kiatt, Mninga, or Mukwa.

This wood is sourced from central South Africa up through East Africa and is closely related to Padauk. The tree is not large, probably only reaching about 21m (70ft) if mature, but can have clear bole up to over 12m (40ft) long. It is not large in diameter, with a maximum of around 60cm (2ft), and lumber tends to be of a smaller specification. When first planed, the wood occasionally has a pleasant, perfume-like odour. The heartwood can be striking, with colours ranging from a mild golden brown through to a darker, reddish or chocolate brown with random yellow or reddish streaking. Logs tend to produce wood with a golden or reddish hue that can make matching difficult at times. The pale yellow sapwood is sharply defined and is usually about 4cm (1.5in) wide. The grain is not often straight and is usually interlocked. Some occasional evidence of white deposits can be found that may mar a finished surface; the texture is fairly coarse.

Muninga is a medium-weight wood weighing between 16 to 18kg (35 to 40lb) when dry per cubic foot, 17kg (37lb) on average. Some variation will occur depending upon local growing conditions. The wood dries slowly without too much trouble, although thicker stock may need some pre-drying to speed up the kilning process. If stacked on thin stickers it will help to avoid too much surface checking. Shrinkage factors are not great and the timber is considered to be stable once dried. Generally it is not a strong wood, rated below Beech. The heartwood is classified as very durable.

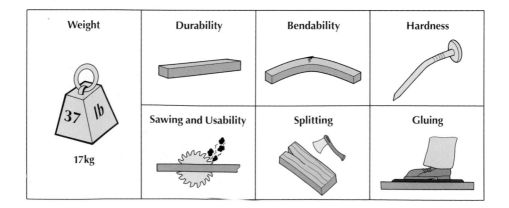

Weight	Durability	Bendability	Hardness
37 lb / 17kg			
	Sawing and Usability	Splitting	Gluing

When a preponderance of uneven and interlocked grain is present, Muninga tends to make working difficult; care, with surface planing especially, will be needed. When cross-cutting it may spelch badly on the back side if not fully supported. With sharp tools it turns well and produces a clean finish. Nailing and screwing operations are best pre-drilled. The white spots occasionally found on finished surfaces can be removed with white spirit. It glues well and a good polished finish can be achieved with care.

This is an attractive timber that is sought after locally for first-class furniture, cabinet and panelling work. It turns well and produces fancy goods including drums and canoes. It is also used as a flooring material, where the colours can be fully appreciated.

Radial

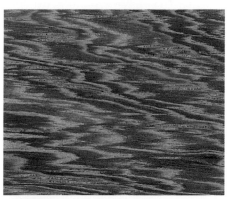

Tangential

Source:
Eastern and Southern Africa.

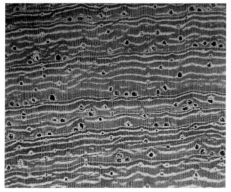

Cross Section

NIANGON

Heritiera utilis

Also known as: Cola Mahogany, Nyankom, Ogoue, Wishmore or Wismore.

This African hardwood is very similar to Mengkulang, which is found in South-east Asia. It generally grows to a medium-sized tree in the countries of West Africa. Initial impressions may lead one to think it is Mahogany, but on closer inspection the texture can be seen to be far too coarse. A fairly wide sapwood is present, slightly paler than the heartwood and not always clearly defined. The heartwood is a pale, light red or pinkish brown in colour with conspicuous dark flecks on the radial faces. The grain is open and coarse with some interlocking present. It feels greasy to the touch and resins can exude from the surface.

When dry, Niangon weighs around 18kg (40lb) on average per cubic foot. It is not particularly strong, with a rating somewhat lower than that of Beech. There is often evidence of brittleheart, shakes along one edge of the planks, which do not help these values. Niangon is a reasonable drier with only a minimal amount of degrade, although some twisting, end splitting and surface checking may occur. Some resin may exude during the drying process; crystallizing on the surface. Providing the timber is kept in stable conditions there will not be much further movement after drying. The heartwood is considered durable and it is extremely resistant to preservative treatment.

When resin is present, some gumming up of a saw's teeth will take place. During cross-cutting operations there may be some tendency for the cuts to spelch out on the back sides.

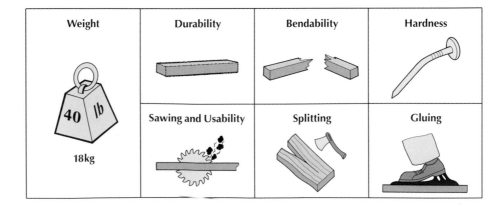

Weight	Durability	Bendability	Hardness
40 lb / 18kg			
	Sawing and Usability	Splitting	Gluing

Apart from these slight problems Niangon is a fairly easy wood to saw and plane, although quarter-sawn faces may tear out somewhat when planing. Because of Niangon's inclination to split, all nailing and screwing jobs should be pre-drilled. The resin content can cause problems when gluing and varnishing, and some surface cleaning with a soluble chemical such as ammonia or caustic soda may be necessary. Prior to staining, the grain will need filling; after this a reasonable finish can be obtained.

Niangon can be used wherever a general quality redwood is required. It is often found in applications where a cheaper substitute for Mahogany is needed, such as furniture, cabinetwork, mouldings and trims. It is also useful for joinery, both internal and external.

Source:
West Africa.

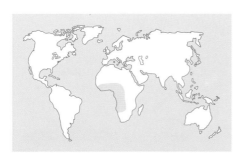

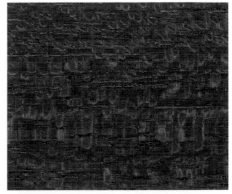

Radial

Tangential

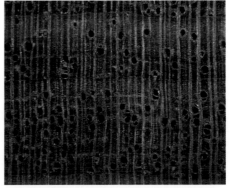

Cross Section

119

NYATOH

Palaquium spp. and *Payena spp.*

Also known as: Njatuh or Padang.

A number of species from Malaysia and Indonesia are put together and marketed collectively. This description focuses on the medium weight group that are generally collected under the main trade name of Nyatoh. Another set of harder, heavier woods from the same species is often marketed as Bitis.

Nyatoh's sapwood is wide, pale and not always distinguishable from the heartwood, which is a deep pale pink through to reddish brown, with darker streaks apparent at times. The grain is of a fairly fine and even nature, but can be somewhat interlocked and slightly wavy. Occasionally on quarter-sawn faces a silky or streaky figure can be seen. It has a dull finish when planed, and overall is not an exciting wood.

The Nyatoh group varies in weight because of the mixture of species. An average dry weight will be around 18 to 21kg (40 to 46lb) per cubic foot, averaging 19kg (42lb). Its strength is rated as comparable to that of Beech. This wood is a fairly slow dryer with some tendency towards surface checking, particularly around any knots, and also to end splits, especially in thicker material. Distortion can also be a problem if not stickered or stacked correctly. It is considered to be non-durable when in contact with the ground, but in other applications is moderately durable; it resists treatment with preservatives.

Nyatoh's working properties are not helped by the presence of silica in some of the collected species. When this occurs blunting of tools can be fairly rapid; otherwise the timber will plane easily and smoothly. A reduction in cutting angle may help

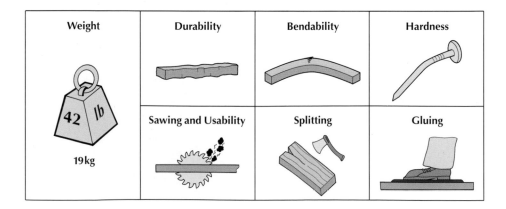

Weight	Durability	Bendability	Hardness
42 lb / 19kg	Sawing and Usability	Splitting	Gluing

to alleviate any possible tearing out on quarter-sawn faces. The wood apparently peels well and is easily turned. Resistance to splitting when nailing is average, but pre-drill when nailing near the end grain or screwing. It glues well, takes stains readily and can be polished to a good finish.

Because of the mix of species, Nyatoh is not always consistent in quality, weight and working properties. As it is a particularly durable wood, it is best used in applications where it will not be in contact with the ground or in variable, moist conditions. It is employed in interior construction in stair parts, joinery, doors and window frames. *Payena spp.* are exploited for 'Gutta Percha' a rubbery substance obtained by ringing the trees. This has been used in the production of moulds in dentistry and for containers of highly corrosive acids.

Source:

South-east Asia.

Radial

Tangential

Cross Section

OAK, AMERICAN RED
Quercus spp.

Also known as: Black, Cherrybark, Laurel, Northern, Northern Red, Nuttall, Pin, Scarlet, Shumard, Southern Red, Spanish, Swamp Red, Water and Willow Oak.

The above listing of 'American Red Oak' contains a number of different species that grow abundantly along the eastern half of North America. Of the two Oak groupings from North America, this one is considered inferior because of its lighter weight, lack of durability and open, coarse texture. It has a pale cream or nearly white sapwood about 2.5 to 5cm (1 to 2in) wide; the heartwood has a distinct reddish brown tinge. There is some considerable variation in the visual appearance of this wood because of the large grouping, and variety of soil and climatic conditions. In general, the grain has an open appearance with the growth rings clearly visible; overall, the grain is relatively straight. There is little evidence of figure on the quarter-sawn faces, although it can be found there if you look for it.

American Red Oak is extremely porous due to the lack of tyloses, a foam-like structure that blocks the vessels of the heartwood. This absence of tyloses can be used to help differentiate between White and Red Oaks, if you smoke! Take a sample about two or three inches long, inhale some smoke, bring the sample to your mouth and try to puff smoke through the length. If it goes through, there is a sporting chance that the wood is Red Oak; if not, it is probably White!

Dry weights of this Oak can vary, but on average are around 18kg (40lb) per cubic foot. It dries slowly with a large shrinkage factor. It may split and check; if care is not taken it

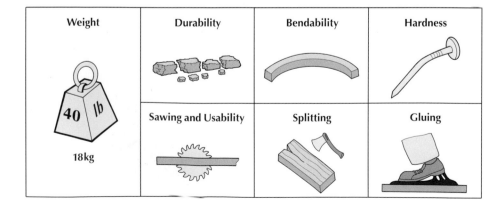

Weight	Durability	Bendability	Hardness
40 lb / **18kg**			
	Sawing and Usability	Splitting	Gluing

can also honeycomb. Rated slightly lower in strength than Beech, it does have an ability to bend well while maintaining good shock resistance. It is classified as non-durable but with only slight resistance to preservative treatment.

When sawing American Red Oak there can be a tendency to bind if the kilning has not been carried out correctly. A good planed finish can be achieved. Pre-drill with all fixings; it glues reasonably well. The surface grain may need some filling before staining and polishing.

American Red Oak, with selection and care, can be used in similar applications to other Oaks; the exception is cooperage. It is best utilized for interior work such as furniture, cabinet work, panelling, trims and mouldings. It has found a niche for kitchen cabinets and doors.

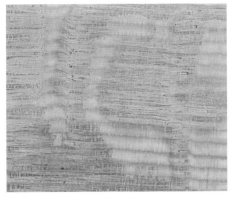

Radial

Tangential

Source:

North America.

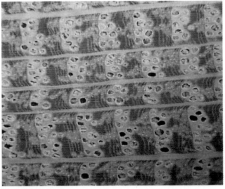

Cross Section

OAK, AMERICAN WHITE
Quercus spp.

Also known as: Bur, Chestnut, Chinkapin, Live Oak, Overcup, Post, Swamp, Swamp Chestnut, Swamp White and True White Oak.

Another large grouping of Oaks from the eastern part of North America, this set is considered superior to the Red assortment. It has a white or pale cream sapwood about 2.5cm (1in) or so wide that is distinguishable from the often variable light to mid-brown heartwood. Similar to its European cousins, but slightly less pronounced, is a silvery figure visible on the quarter- sawn faces. Depending upon source the structure and appearance can be quite variable. With both fast and slow grown trees the growth rings are clearly visible and the grain is fairly open; small knots may also be apparent. Tyloses is present and reflects light at certain angles. This wood does not like ferrous metals and will stain dark brown or black in moist conditions if contaminated.

Just slightly heavier than Red Oaks, White Oak weighs up to an average of no more than 20kg (44lb) per cubic foot when dry. Without as much shrinkage as Red, degrade during drying tends to be from cupping, checking and splitting, with some tendency towards honeycombing if care is not taken. Drying thicker than 7.5cm (3in) stock is not recommended because of the time it takes and the greater chance of degrade. White Oak is rated as just slightly lower in strength values to Beech. It has good steam-bending properties, but make sure it is kept away from anything that might stain it. The heartwood is considered durable and is extremely resistant to preservative treatment due to the presence of tyloses.

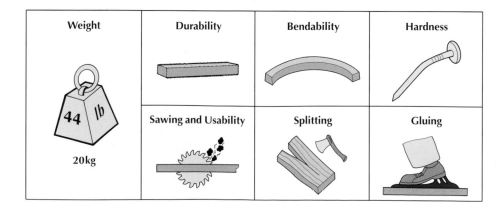

Weight	Durability	Bendability	Hardness
44 lb / 20kg			
	Sawing and Usability	Splitting	Gluing

Kiln-dried wood may distort off the saw when cut if care has not been taken during that process. White Oak is slightly milder than European, therefore it is just a bit easier to plane. On occasions, a slightly reduced cutting angle may produce a better finished surface. Pre-drill before nailing or screwing. Gluing can be variable. The grain will need filling before staining and polishing; an excellent finish can be obtained.

Because of the variety of species, careful choice of stock is important. The inclusion of tyloses makes this an ideal timber for cooperage purposes, particularly whisky casks. Its strength and durability allow use for constructional purposes and all types of joinery. It makes an ideal replacement for European Oak, especially in furniture.

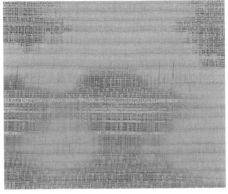

Radial

Tangential

Source:

North America.

Cross Section

OAK, TASMANIAN

Eucalyptus delegatensis

Also known as: Gum-Top or White-Top Stringybark, Alpine or Red Mountain Ash or Woolly Butt.

(*E. obliqua*)
Also known as: Messmate, Stringybark Messmate, Brown-Top Stringybark or Stringybark.

(*E. regnans*)
Also known as: Swamp Gum, Mountain or White Ash.

Quite a mixture as you can see! There are three main species of Eucalypt that are collectively marketed and sold overseas as Tasmanian Oak. They may also be called Australian or Victorian Oak. These timbers are individually important in Australia for their commercial values. *E. regnans* is considered to be possibly one of the largest hardwood trees in the world.

Obviously not true Oaks, or Ashes for that matter, they have a very loose and general resemblance to plain-sawn Oaks. There is no display of any figure or silver grain on the quarter-sawn faces. With well-defined growth rings, Tasmanian Oak typically has a coarse, fairly straight grain that can include some gum veins.

Weight, when dry, can be varied, but an average of around 21kg (46lb) per cubic foot seems about right. Strength values are slightly better than those of Beech. Drying is fairly rapid with significant shrinkage. There is a tendency to degrade through surface checking early on in the process. Some distortion and case-hardening may also be a problem. Checking and case-hardening can both be reduced by conditioning at the end of the kiln run. The heartwood is moderately

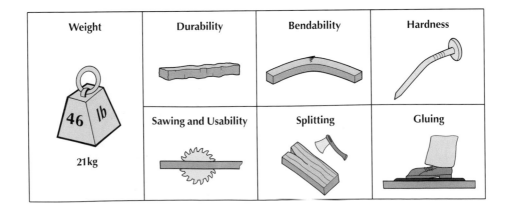

durable and resists treatment with preservatives.

Tasmanian Oak is fairly easy to saw and plane, with only occasional resharpening required. Steam-bending can produce variable results depending upon which of the Eucalypts is being used. The harder species tend to split when nailed, so it is a good idea to pre-drill before all jointing functions. Gluing is satisfactory and stains can be applied evenly. With effort, a nicely polished surface can be achieved.

This group of species is collectively and individually important to Australia. The slightly heavier *E. obliqua* is used for structural work, including posts, piles, wharf and dock works and railway sleepers. The other two are used in lighter work. The group as a whole is a good all round source of timber that can be used for many applications.

Source:

Southern Australia.

Radial

Tangential

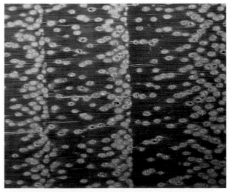

Cross Section

129

OBECHE

Triplochiton scleroxylon

Also known as: African Whitewood, Arere, Ayous, Egin-fifen, Ewowo, Obechi, Okpo, Owowa, Samba, Samba-ayous and Wawa.

This is one of Africa's larger forest trees, reaching heights of up to 60m (200ft) with clear boles that can be more than 1m (3 to 4ft) in diameter above the buttress. Although widespread, the main growing areas are the countries of West Africa. If not called Obeche, it will most likely be known as Wawa, a name attributed to the wood from Ghana. It has a very wide sapwood that leads into the heartwood without any noticeable differentiation. The heartwood is a light creamy white or pale straw colour darkening slightly on exposure. A lightweight timber, the grain is usually uniform, typically interlocked and of a fairly open nature. Plain when flat sawn,

Obeche's grain configuration generally leads to an attractive, pronounced striped effect being apparent on the quarter-sawn faces.

This is one of the lightest-weight commercial timbers available. Its dry weight average is around 11kg (24lb) per cubic foot. It is often used as another of the benchmarks against which to measure comparative strength values. Although the strength to weight ratio is good, it is not in fact a strong wood and use in areas of structural importance should be avoided. It is quite resistant to impact but will split easily. From the log, sawn lumber should be stacked on stickers immediately to avoid any sap or surface staining due to fungal attack. It is an extremely quick dryer with a reputation for little shrinkage or degrade during this process. As one would expect, Obeche is not

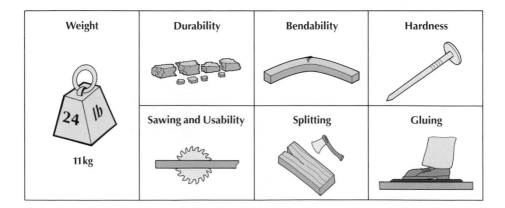

Weight	Durability	Bendability	Hardness
24 lb / 11kg	Sawing and Usability	Splitting	Gluing

rated as durable and rarely finds a suitable external use.

Obeche's softness makes it a fairly easy timber to work. To maintain a good finish, all tools should be kept sharp. It does have a tendency to crumble when working on the end grain. It nails and screws well without too much splitting, but will not hold fixings under pressure – add glue to be sure. Its open-grain structure accepts glue well but does mean that a certain amount of filling is required before finishing. It stains well, but watch absorbency rates near the end grain; can be polished well.

Because of a similar grain appearance to Mahogany it has been stained and used as a replacement. It is most commonly found as a foundation framing wood in furniture, although it also peels well and is often used as a core wood in plywood manufacture.

Source:

West Africa.

Radial

Tangential

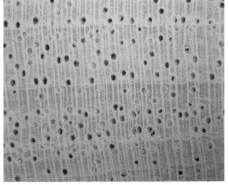

Cross Section

OPEPE

Nauclea diderrichii (formerly *Sarcocephalus diderrichii*)

Also known as: Akondoc, Alama, Badi, Belinga, Bonkangu, Bonkese, Ekusawa, Gulu, Kantate, Kusia, Kusiaba, Kusiabo, Maza, Ngulu, Obiache or Ubulu.

Principally commercially cropped throughout the countries of West Africa, this tree also grows in the central region. Another strong wood, the Opepe tree grows up to around 46m (150ft) tall, producing clear bole lengths at about a maximum of 24 to 30m (80 to 100ft). The sapwood is a distinct pale yellowish or pinkish colour. In contrast, the heartwood starts out as an orange-brown or golden yellow. This darkens on exposure to a bright, clean, rich golden brown with some occasional darker streaks. The grain is medium to coarse in texture and is slightly interlocked or irregular; this produces a mild, attractive stripy figure.

Opepe is a hard and reasonably heavy wood at around 21kg (46lb) on average per cubic foot. Its strength is considered exceptional and is rated above that of Beech; straight-grained material is considerably more so. It is not a good timber to bend. Drying quarter-sawn Opepe is relatively quick and easy but can be traumatic if a large proportion of the lumber is flat sawn. In the latter instance, a great deal of surface checking may take place; coupled with some distortion downgrade volumes can be high. A slower kilning rate is preferable and some pre-drying on thin stickers is thoroughly recommended. The heartwood is considered very durable and moderately resistant to preservative treatment.

Opepe has reasonable working qualities with little occurrence of blunting to tool edges. When planing,

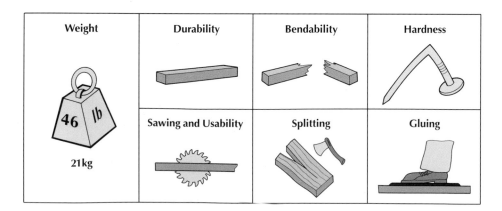

Weight	Durability	Bendability	Hardness
46 lb 21kg			
	Sawing and Usability	Splitting	Gluing

a much reduced cutting angle will help to avoid the surface from picking up or tearing out. With a tendency to split, it is good practice to pre-drill when nailing and screwing to avoid disappointment! It glues well, taking advantage of the open grain structure. Fill the wood before staining and polishing to produce the best results. This is a nice wood to finish with exceptionally good polishing properties.

This strong, durable and versatile wood has found the usual traditional uses for pier and dock work, decking and piling. Its tendency to surface check has restricted it from areas where there is a likelihood of variable conditions. If sealed and surface treated correctly, it can make an excellent exterior joinery material. Hard-wearing properties have also lead to a use for flooring.

Radial

Tangential

Source:

West and Central Africa.

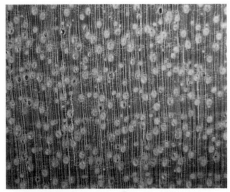

Cross Section

133

PADAUK

Pterocarpus spp.

Also known as: African, Andaman and Burma Padauk, African Coralwood, Amboyna, Andaman Redwood, Barwood, Camwood, Chanlanga-da, Corail, Mai Pradoo, M'bel, Mbeu, Muenge, Pradoo or Vermilion Wood.

Some timbers of a grouping of this species are collectively called Padauk. They have common features and are here reviewed together with some explanations. There were four principal sources: West Africa, Myanmar (Burma), the Andaman Islands, a group of islands between India and Malaya in the Indian Ocean, and the Moluccas group of islands found in South-eastern Indonesia. These latter two sources probably produce the most prized timbers. The Indonesian island of Amboyna, or Amboina, was once responsible for the decorative wood that is separated and called by the name 'Amboyna'. This is a trade name that is applied to timber produced from burr growth on the trees most commonly found on *P. indicus*, but also on other members of the species. It produces a burr or bird's eye type of wood that is sought after as a veneer. 'Camwood' is a misuse of a name that should be applied to another West African hardwood, *Baphia nitida*. The true Camwood is in short supply and therefore Padauk has been used as a substitute; to add further confusion, both are used to extract a natural dye by boiling wood chips in water!

Padauk, in this collective assessment, produces a striking blood red to brown heartwood with some dark streaking; all darkens to a more uniform purple brown on exposure. The sapwood is a distinctly paler colour. The grain is generally coarse

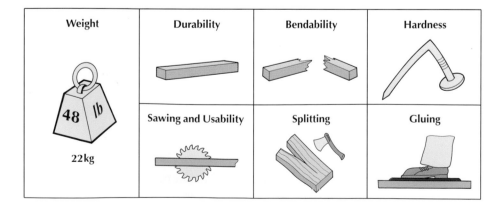

Weight	Durability	Bendability	Hardness
48 lb 22kg			
	Sawing and Usability	Splitting	Gluing

in texture with an amount of interlocking.

There can be some variations in weight but generally, when dry, it will range from 21 to 23kg per cubic foot, averaging 22kg (48lb). When available, Burma Padauk tends to be slightly heavier. This is a mid to heavyweight timber with a strength rating similar to Beech. It dries reasonably well but slowly with some checking, but is considered stable. It is another tree that is often 'girdled'. Heartwood is classified as very durable.

Padauk works well apart from some slight tearing during planing; it turns well. Pre-drill for all nailing and screwing. It glues well; some surface finishing may be necessary before polishing.

This wood is prized for its colour, strength and stability.

Radial

Tangential

Source:

Africa through to South-east Asia.

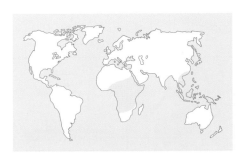

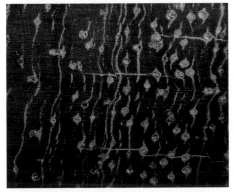

Cross Section

PEAR

Pyrus communis

The Pear tree is grown principally for its fruit which, depending upon variety, will produce an edible pear for using in jams and preserves, or a dry 'perry' pear that is used to make a drink similar to cider. Large batches of timber are not normally available; most will come from old orchards. It is another of the country timbers that can be called 'fruitwood', a collective name used along with apple, plum and cherry. When freshly sawn, the heartwood of Pear is a light pinkish brown with a definite purple hue. This evens out to a slightly darker reddish brown on exposure. There is no sharp difference between the sap and heartwood although it is discernible if looked for. The grain is tight, fine and evenly textured. Because of the likely growing circumstances there will be an amount of interlocked and irregular grain; some spiral grain can also occur.

Pear is just a bit heavier than Apple, on average about 21kg (46lb) per cubic foot when dry. Its tight and even grain texture plus weight give it the feel of a hard wood; it is fairly tough with a resistance to splitting. It dries slowly and will show a distinct inclination to distort. This is because it is usually grown in orchards or other open areas, creating the interlocked and spiral grain configurations. It is not a durable timber and should always be treated for external use. It is susceptible to attack by furniture beetle under favourable conditions.

Pear is relatively easy to work with both saw and plane, except in cases where the grain is particularly wild. It produces an interesting 'tube' like shaving when machine planed. It is an

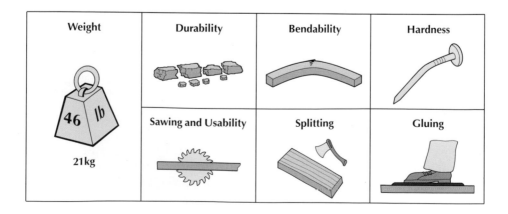

Weight	Durability	Bendability	Hardness
46 lb / 21kg	Sawing and Usability	Splitting	Gluing

excellent wood to carve when dry, with the added advantage of being reasonably stable; it turns well also. Although it has a reputation for not splitting it is still advisable to pre-drill before fixing with nails or screws. It will glue well and needs little preparation before staining and polishing.

When available, this wood can make some fine furniture. It has found uses for musical instrument parts, drawing instruments, fancy turnery and household tableware. It was traditionally used, with other fruitwoods, for cottage furniture and internal joinery such as doors and cupboards. It is also popular as a carving medium, and also for some inlay work.

Radial

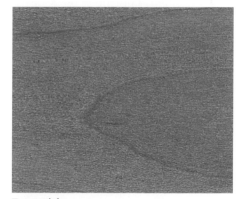

Tangential

Source:

Throughout Europe and similar climates.

Cross Section

PLANE, EUROPEAN

Platanus orientalis and hybrida, (formerly *P. acerifolia*)

Also known as: English, French, London Plane, and so on, depending upon source, and as Lacewood.

There are two species growing in Europe and on into Asia Minor. *P. orientalis* is to be found in south-eastern areas and *P. hybrida* in central and northern Europe. This latter one is generally considered to be a hybrid between *P. orientalis* and *P. occidentalis*, a large tree from the North American continent. To confuse matters further, in Scotland and northern parts of the UK the name Plane is also attributed to the Sycamore tree! Not normally grown for its commercial value, the Plane tree is to be seen lining the streets of London, Paris and other capitals of Europe. It takes kindly to most climates and will also stand a fair bit of pruning. When wood becomes available there is little to tell the difference between the sap and heartwood. The heartwood will bear some resemblance to Beech in colour, being a pale creamy brown, often with a dark core. When cut on the true quarter the rays are clearly seen as a 'chink'-like figure; it is wood from these cuts that is known as Lacewood. This is prized as a useful timber from which to produce decorative veneers.

When dry, Plane weighs around 17kg (37lb) on average per cubic foot. It has a medium hardness that is rated as somewhat weaker than Beech. A quick dryer with a tendency to warp, there will also be relatively more shrinkage tangentially than radially. This is not a durable wood and should not be used in applications likely to expose it to varying moisture conditions. When used in furniture, under favourable conditions it is susceptible to attack by the furniture beetle.

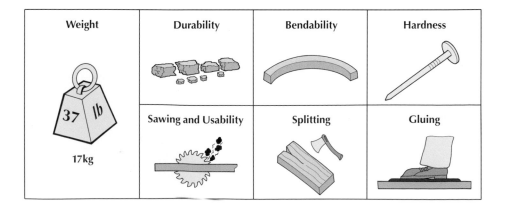

Weight	Durability	Bendability	Hardness
37 lb / 17kg			
	Sawing and Usability	Splitting	Gluing

This not a difficult timber to work, with little blunting taking place, although it will have a tendency to bind on the saw. On quarter-sawn material, the rays may flake out, so ensure that all knives are kept sharp. Nailing is satisfactory but pre-drill for screwing. It glues well and will take a stain fairly evenly, but watch out near the end grain. It will polish quite well with a bit of effort.

Plane is mostly prized for its decorative properties. It is used to make small boxes and chests, and for inlay work, veneers and similar applications. Lacewood, in particular, is sought after in wider boards for panelling.

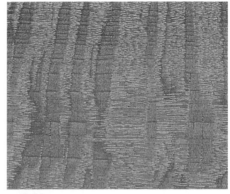

Radial

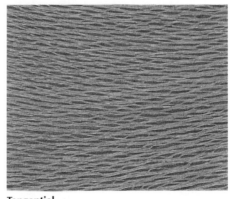

Tangential

Source:

Throughout Europe and similar climates.

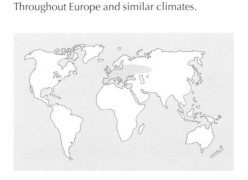

Cross Section

POPLAR

Poplus spp.

Also known as: Aspen and European Aspen, Cottonwood and Black, Grey, Italian Black, Lombardy Poplar.

The Poplar tree, in various forms, is widely dispersed throughout Europe and North America. Because of a similarity of properties between them the varieties have, for this description, been placed together. Poplars are medium sized trees, with the tallest rarely exceeding much over 38m (125ft) in height. Most commercial lumber is produced from Aspen and Cottonwood grown in North America and the Black Poplars in Europe. The timber shows little difference between the sap and heartwood; both are creamy white through to a light reddish brown. There can be some streaking accentuating the grain pattern, with a darkening of the wood towards the core. The grain is usually straight with a fine and even texture that typically leans towards a woolly surface when cut. Some evidence of small 'pin' knots can often be seen on planed faces.

This is a lightweight wood that when dry will average about 13kg (29lb) per cubic foot. Strength to weight ratios are good and it is considered quite a tough timber, although rated well below Beech. It dries fairly easily and rapidly, but has a tendency to split around any knot holes. Some uneven drying may take place that results in pockets of higher moisture content within individual planks. It is not durable and any applications that may put it under stress will cause it to rot very quickly; always treat with preservatives for outside use.

The woolly nature of this timber will find saws binding slightly in use.

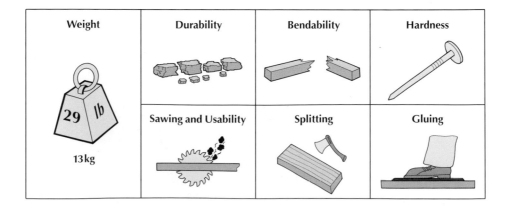

This can also impact upon the planed surface finish; reducing the angle of sharpness and keeping all edges keen will help to alleviate this. The wood is fairly brittle and therefore does not bend well, although it does have a fair resistance to splitting. This is a good timber to peel for use in plywood manufacture. It nails and screws satisfactorily and will also glue well. Watch out if staining, as results can be variable. It will polish to a reasonable finish once any woollyness has been dealt with.

Poorer quality material is used up for pallets and other similar utility jobs. It is the main timber from which matchsticks are produced, and, thinly sliced, traditional matchboxes. It is a good material as a core element in plywood thanks to its weight and toughness. It is also an important source of pulpwood.

Radial

Tangential

Source:

Throughout Europe, North America and similar climates.

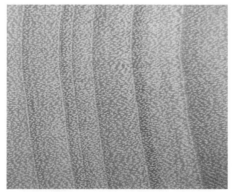

Cross Section

PURPLEHEART

Peltogyne spp.

Also known as: Amarante, Amaranth, Bois Poutpre, Coataquiana, Kooroobooelli, Palo Morado, Palo Nazareno, Pau Ferro, Pau Roxo, Purperheart, Purplewood, Sakavalli, or Violet Wood.

The trees from which the Purpleheart wood comes are to be found throughout Central and South America. Depending upon local conditions they can grow to between 38 to 45m (125 to 150ft) high. With fairly straight boles up to 15m (50ft) long before branching, diameters can be up to 1.2m (4ft) or so. When cut, the sapwood, a pale grey or dirty white, is clearly distinguishable from the heartwood, which when first cut, is not striking, being a browny olive colour; however, after a while this turns into a bright, vivid purple. Unfortunately this does not last when left in direct sunlight and soon turns a rich browny red; cutting a thin slice from this surface will once more reveal the purple undercolour. Some of the species will retain their colour better than others. When any are used internally, the toning down stage takes longer and may remain for some time. The grain is generally straight but does incline towards being irregular or interlocking; this produces some striping on the quarter cut. It is fairly fine to medium in texture.

This is a heavy wood, at anything from 25 to 29kg (55 to 64lb), per cubic foot, averaging 27kg (60lb). It is sought after for its strength properties – it is rated much better than Beech but not quite up to Greenheart. It has extremely good impact resistance, and is a moderate bender. It dries at a steady rate with only a small amount of surface checking, although on occasions it can be difficult to remove

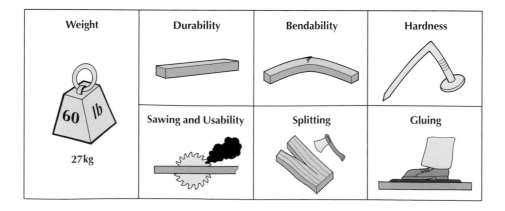

Weight	Durability	Bendability	Hardness
60 lb 27kg			
	Sawing and Usability	Splitting	Gluing

some pockets of moisture. It is rated as very durable and does not accept preservatives easily.

This is a difficult wood to saw. It tends to burn and bind, and there can be a build-up of gum on sawteeth. When planing, a slow feed is probably preferable to avoid too much picking up of the grain; always keep tools sharp and possibly reduce the cutting angle. Purpleheart turns well. All nailing and screwing functions will require pre-drilling. It takes glues satisfactorily and can be polished to a fine surface finish.

This is another good heavy duty structural and constructional timber that finds many suitable applications in its source countries. It is occasionally used for decorative inlay work on furniture, and often for fancy turned goods, flooring and tool handles.

Radial

Tangential

Source:

Central and South America.

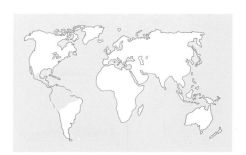

Cross Section

RAMIN

Gonystylus spp. (Normally G. bancanus).

Also known as: Ahmin, Melawis or Ramin Telur.

Although members of the Gonystylus species are found from India to the Philippines, the main commercial source of timber comes from Malaysia and Indonesia, in particular the island of Borneo. The tree is not large but is abundant in coastal swamp forests. Clear boles of 8 to 11m (25 to 35ft) are small in diameter, producing mainly strips and scantlings. Most of the raw material is processed in the source countries; when lumber is available it is rarely thicker than 4 to 5cm (1.5 to 2in) and only up to 15 to 20cm (6 to 8in) wide! The sapwood is not clearly distinguishable from the pale cream or straw coloured heartwood. The grain is medium to fine and usually straight and even; there are few distinguishing features apart from a

slight musky smell. Because freshly sawn lumber is very susceptible to attack by blue stain fungi, the wood is usually dipped or sprayed with an anti-stain treatment immediately after conversion. Splinters from timber so treated need to be avoided, as they can cause minor, short-term infection.

Ramin is a medium to lightweight timber, averaging around forty to 19kg (42lb) per cubic foot when dry. It has a comparable strength rating to that of Beech but is not as tough. With lower shear values, it does not measure up as a bending timber. It dries reasonably well and quickly, with little shrinkage, and there is minimal degrade apart from a marked amount of end splitting. Additionally, if care is not taken with those boards over 4cm (1.5in) thick a significant amount of surface checking can occur. All Ramin wood is perishable and is not suitable

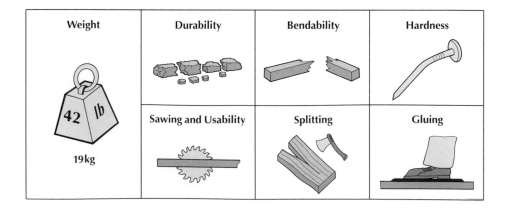

Weight	Durability	Bendability	Hardness
42 lb 19kg			
	Sawing and Usability	Splitting	Gluing

for external use where it will degrade very quickly; it does accept treatment if you wish to take the chance!

This wood has good working properties, with little blunting of tools taking place. Occasionally, material with a refractory grain can prove slightly more difficult to plane. It has a tendency to split, so pre-drill when nailing and screwing. It glues well and accepts stains readily, but watch out for greater absorbency towards the end grain. Little filling is required to produce a good, but rather plain, surface finish.

Ramin is most commonly processed in its source countries to produce small mouldings, dowels, trims and other lightweight items. As a bland and easily stained wood, when available it is used to produce furniture parts.

Radial

Tangential

Source:

Predominantly Malaysia and Indonesia.

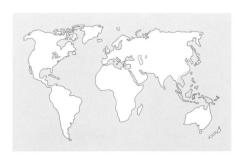

Cross Section

ROSEWOOD, BRAZILIAN

Dalbergia nigra

Also known as: Bahia or Rio Rosewood, Cabiuna, Jacaranda, Palissandre, Pau Preto or Pau Rosa.

Of the three commercial Rosewoods this one, when mature, is probably the larger tree. It is a scarce and expensive resource. When available, it is found in eastern Brazil, mainly around the Rio de Janeiro area, and grows up to 38m (125ft) or thereabouts. The trunk is not large, therefore boards tend to be long and narrow rather than wide. When freshly cut, the timber from mature trees has a distinctive odour that is slightly reminiscent of rose fragrance; if a small sample is sucked it also has a slightly perfumed taste. With wood from immature trees, these features are often not so pronounced nor is the colouring so distinct. The scent and taste come from the natural oils that are present throughout the structure of the wood. Additionally, gums are often to be found in the open, coarse pore structure. When tilted to catch the light, the gum in Brazilian Rosewood is reflected as a deep black colour; in the other Rosewoods this is slightly lighter. The sapwood is a clearly defined pale grey or nearly white colour. The heartwood is anything from a dark orange or chocolate brown through to a violet brown with darker or black streaks. Some of the most prized lumber is produced from old, slow grown, possibly defective logs. The wood naturally feels oily to the touch. The grain can be straight but tends to be irregular, especially in really mature trees; it is coarse in appearance and slightly interlocked.

Along with Indian Rosewood, Brazilian Rosewood is slightly lighter than the Honduras Rosewood,

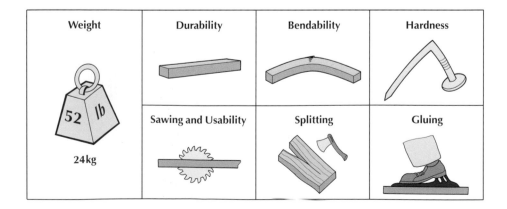

Weight	Durability	Bendability	Hardness
52 lb			
24kg	Sawing and Usability	Splitting	Gluing

although it is still hard and heavy at around 23 to 25kg (52 to 55lb) per cubic foot, averaging 24kg (52lb). It is a slow dryer and care should be taken to ensure surface checking is minimized. This is a strong and naturally very durable timber.

Although hard, Brazilian Rosewood is not difficult to work and can be planed to a smooth finish. It will bend with care, all nailing and screwing operations should be pre-drilled. The presence of natural oils makes gluing trying at times and producing a high polish can be problematic.

Brazilian Rosewood has been commercially exploited for over three hundred years. It is prized for high-class cabinet work and for musical instruments. The use of solid wood today has usually been replaced by veneering.

Radial

Tangential

Source:

South America, Brazil.

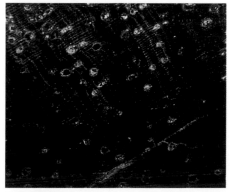

Cross Section

ROSEWOOD, HONDURAS

Dalbergia stevensonii

Also known as: Nagaed.

This tree from Central America is not as tall as its Brazilian cousin but is quite a bit heavier. Although principally available in small commercial quantities from Honduras, it is found throughout Central America. A mature tree will reach heights of 23 to 30m (75 to 100ft) but will probably fork at around 8 to 9m (26 to 30ft). Lumber is not particularly long or wide. The wood will have the distinctive odour of rose fragrance when first cut, from the natural oils present within the timber, but unlike the Brazilian Rosewood it has no taste. However, as with the Brazilian it has gum deposits in the open grain structure that will reflect the light when tilted, creating a slightly golden hue. The sapwood is fairly narrow at around 2.5cm (1in) wide. It is clearly distinct and starts out

as an off-white colour that quickly turns a yellowy cream on exposure. The heartwood is similar to the other Rosewoods in appearance.

This is the heaviest of the Rosewoods, ranging between 25 to 29kg (55 to 64lb) per cubic foot when dry, with an average of 27kg (59lb). Also a slow dryer, it tends to surface check quite badly if care is not taken; stacking on thin stickers initially is highly recommended. Once dry, it is considered to be a very stable timber. It is strong and hard and the heartwood is very durable, although these properties are not often exploited in its end uses.

Probably the most difficult of the Rosewoods to work, blunting of tools will surely take place. Its hardness leads to some chattering if not held firmly with pressures when planing, although a smooth planed finish can

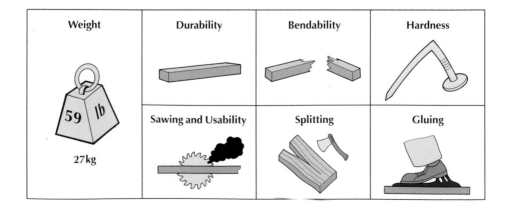

Weight	Durability	Bendability	Hardness
59 lb / 27kg			
	Sawing and Usability	Splitting	Gluing

be achieved with care. This is a good turning wood so long as tools are kept sharp. It is not easy to bend and requires pre-drilling before nailing or screwing. The natural oils present a problem again when gluing. They also impact upon the surface quality, with polishing becoming difficult unless the wood is sealed correctly.

Honduras Rosewood shares many of the properties of Brazilian Rosewood and has therefore been used for similar purposes as a substitute as well as in its own right. The logs and lumber are not available in large sizes and this is refelected in its end uses. It is a popular wood for high-class cabinet work, internal panelling and musical instrument keys.

Radial

Tangential

Source:

Central America, Honduras.

Cross Section

ROSEWOOD, INDIAN

Dalbergia latifolia

Also known as: Angsana Keling, Blackwood, Bombay Blackwood, East Indian Rosewood, Java Palissandre, Shisham or Sonokelin.

Never particularly abundant, this Rosewood is found throughout most of India apart from the north-west. It also grows and produces commercial timber in Indonesia, especially on the island of Java. Similar in height to Honduras Rosewood, the tree will vary according to local growing conditions. The boles tend to be clear of branches up to anything from 8 to 15m (26 to 50ft) and are generally straight and round. Like the other Rosewoods, this wood has the fragrance of roses when freshly cut. It is likewise oily to the touch and can have gums or occasional mineral deposits in the pore structure. The sapwood is clearly distinct and is a creamy white to off-yellow colour

sometimes with a flush of pink. The heartwood is very similar to the other Rosewoods in colour and appearance. The grain is open and moderately interlocked, with a medium to coarse texture.

It is similar in dry weight to Brazilian Rosewood at about 23 to 25kg (50 to 55lb), 24kg (52lb) on average, per cubic foot. It is particularly hard and is rated as slightly stronger than Beech. Indian Rosewood dries fairly rapidly, which can cause problems with surface checks and end splits. If any are present they will probably get longer and wider as the process proceeds. Stacking on thin stickers to slow down the air circulation will help to reduce the problem. Shrinkage is not particularly noticeable. The heartwood is naturally very durable and the wood is reputed to survive for

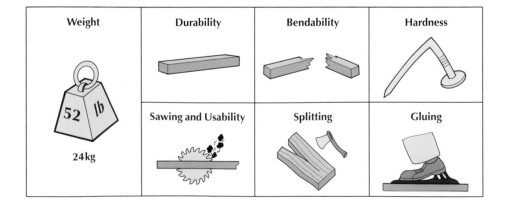

Weight	Durability	Bendability	Hardness
52 lb 24kg			
	Sawing and Usability	Splitting	Gluing

long periods if totally immersed in water. However, none of these properties are reflected in its end uses because it is too scarce and expensive!

The working qualities of Indian Rosewood are nearly as difficult as those of its cousin from Honduras. The occasional presence of minerals increases the blunting effect on saws and planes, therefore tools need to be kept sharp. It turns well and a smooth finish can be achieved with this and during planing; once more care needs to be taken with gluing due to the natural oils. Pre-drill for nailing and screwing. With care and effort, a good finish can be achieved providing the wood is sealed correctly.

Like the other Rosewoods, this is a highly decorative wood. It is sought after for furniture, cabinetwork and veneers. A lot of smaller turned items are produced from it and it is used for musical instruments.

Source:

India and parts of Indonesia.

Radial

Tangential

Cross Section

SAPELE

Entandraphragma cylindricum

Also known as: Aboudokro, Acajou Sapelli, Botsife, Liboyo, Lifaki, Penkwa, Sapele Mahogany or Scented Mahogany or Sapelli.

Sapele is found from West through Central to East Africa; the bulk of the commercial timber production comes from West Africa. It is from the same family, *Meliaceae*, as Mahogany and can often be confused with African Mahogany. The tree is tall at around 46m (150ft) with slightly buttressed clear boles. These can be over 30m (100ft) long with diameters up to 1m (3 to 4ft). It produces fairly long and wide lumber that is usually cut into random widths and lengths. The sapwood is a pale yellow or creamy white that is clearly distinct from the medium red to slightly brown heartwood. The grain is slightly interlocked, which produces the quarter-sawn stripe figure for which Sapele is renowned. In addition to this figure, it can have an attractive fiddleback or roe-like appearance. The texture of the grain is fairly fine, although it can vary widely depending upon local growing conditions. Sapele is very difficult to tell apart from its nearest cousin, Utile, normally microscopic analysis is necessary to determine which is which. However, those practised in handling the raw lumber of both claim to be able to tell the difference by the spicy, cedar-like smell that is associated with Sapele when first cut. Sapele may also be slightly lighter in colour when compared with Utile but, again, local growing conditions will affect this.

When dry, Sapele will generally weigh around 17 to 19kg (37 to 42lb) per cubic foot, on average 18kg (40lb). It is a fairly tough, hard timber

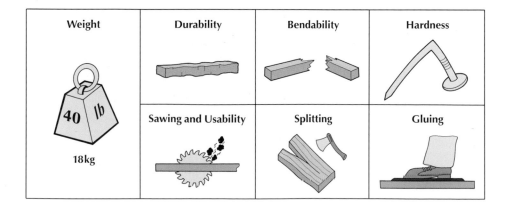

Weight	Durability	Bendability	Hardness
40 lb **18kg**			
	Sawing and Usability	Splitting	Gluing

that equals Beech in strength values; it does not bend well. It dries fairly rapidly but has a reputation for distortion; this can be minimized by cutting and using timber as close to quarter sawn as possible. The heartwood is considered to be moderately durable.

Sapele can be difficult to saw, with some binding and burning if the grain is particularly interlocked. A good planed surface can be achieved with care. It turns well, but has a tendency to split, so pre-drill all fixings. It will glue well, take stains if necessary and will polish to a fine finish.

A good wood for furniture and joinery work, it is particularly well known for its use as a veneer for door facings. When used as an interior flooring material, Sapele can be very attractive.

Radial

Tangential

Source:

Africa, predominantly West Africa.

Cross Section

SEPETIR

Sindora spp.

Also known as: Gu, Makata, Petir, Sampar Hantu, Sindur, Supa or Tampar Hantu.

The timber from Sepetir should not be confused with that of 'Swamp Sepetir' (*Pseudosindora palustris*), which is also occasionally marketed under the same name. The tree is found throughout South-east Asia, with principal supplies coming from Malaysia and Indonesia. It is a large tree, growing up to 46m (150ft) high. Cut boles are usually anything from 6 to 15m (20 to 50ft) long and can be as wide as 1.2m (4ft) in diameter although they are more typically around 1m (3ft). The sapwood can vary wildly in width, and is a dull greyish brown with some occasional pink streaks. The heartwood is clearly visible as a golden red-brown that darkens on exposure; it is often streaked with darker brown. The grain is fairly straight or shallowly interlocked; its texture is fine and even. Some striped figure can be found on quarter-sawn faces. The wood feels slightly greasy to the touch. Resins are present and the wood has a light, spicy smell when freshly sawn.

This is a medium-weight timber. Although there can be some variations, the dry weight average is 19kg (42lb) per cubic foot. Similar in strength to Beech, it has reasonably good resistance against shearing and is fairly stiff; it is not a good bender. It dries fairly slowly with little shrinkage but has a slight tendency to distort and end split. The heartwood is considered durable but the sapwood is perishable which may cause problems because of its width which can be up to 30cm (12in).

The presence of resins tends to make sawing and planing difficult due

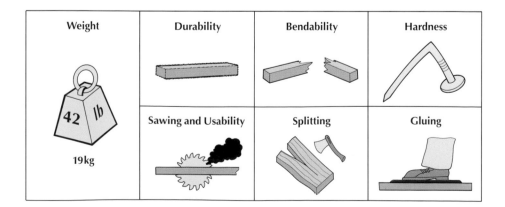

Weight	Durability	Bendability	Hardness
42 lb / 19kg			
	Sawing and Usability	Splitting	Gluing

to some gumming up of the tools; they will need to be kept clean to ensure a reasonable finish. Care needs to be taken when planing to avoid any surface pick-up or tearing out. It splits easily, therefore pre-drill when nailing and screwing to avoid disappointment. It glues reasonably well, takes stains if necessary and can be polished to a satisfactory finish.

The darker heartwood of Sepetir can be attractive when used in furniture and cabinet work; its stability after drying can also be advantageous for these uses. It is suitable for joinery but avoid any sapwood. Some use has been found for it in sporting goods, panelling and musical instruments. It is often used as veneers in local production.

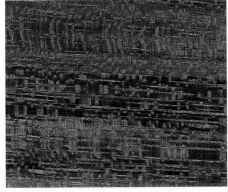

Radial

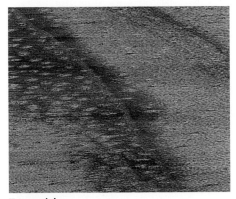

Tangential

Source:

South East Asia.

Cross Section

SNAKEWOOD

Brosimum spp.

Also known as: Cardinal Wood, Conouru, Leopard Wood, Letterhoot, Letterwood or Tortoiseshell Wood.

This is a rare, small tree, not much more than 24m (80ft) when mature, growing in the north-eastern parts of South America. Only short, clear boles of up to a maximum of 60cm (2ft) diameter, but generally less, are occasionally available. Most of the limited amounts of Snakewood that do become available are sourced from French Guyana and Surinam. It is a hard and heavy wood, sought after for its figured heartwood. Often the sapwood is cut off at source and the boles and billets are sold by weight. When attached, the sapwood is clearly defined as a creamy white colour. The heartwood usually has a deep reddish brown base with irregular darker brown or black vertical markings running across the grain radially.

These darker stripes run through the wood in a broken fashion, leading to spots and blotches occurring across every surface; the overall effect can be stunning! The grain is usually straight or shallowly interlocked and of a fine and even texture; the wood feels cold to the touch.

Snakewood is one of the hardest and heaviest woods on the market. When dried, it will weigh anything from 34 to 39kg (75 to 85lb) per cubic foot, on average. This is a difficult timber to dry. There is usually a preponderance of surface checking and a fair amount of warping. As is usual with harder, heavier woods, it is advisable to do a fair amount of pre-drying before kilning. Stack on very thin stickers, to slow the air circulation, and try weighing down the top of the stack to help to reduce warping. When eventually dry, this

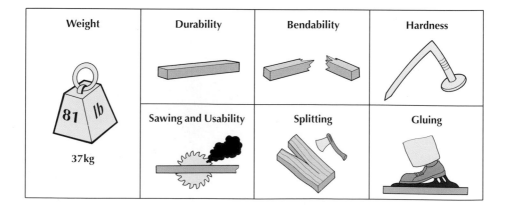

Weight	Durability	Bendability	Hardness
81 lb 37kg			
	Sawing and Usability	Splitting	Gluing

wood is fairly stable. Although heavy, Snakewood is somewhat brittle and also tends to split easily. The heartwood is naturally very durable.

Working with Snakewood is not easy! Tools soon lose their edge and need constant sharpening. It slices well for veneers and turns to a good, clean finish; pre-drill for all fixings. Some care needs to be taken when gluing. When a satisfactory surface finish has been achieved, it can be highly polished.

Snakewood's rarity has made it a sought-after timber for smaller items such as walking sticks, fancy turnery, snooker and pool cue butts, inlay work and violin bows. The best of the figured wood will fetch high prices.

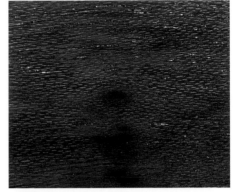

Radial

Tangential

Source:
South America.

Cross Section

SHOREAS, DARK RED

Shorea spp.

Most commonly known as: Dark Red Meranti, Lauan and Seraya.

Trees of the Shorea species grow prolifically throughout the countries of South-east Asia and are known widely for their commercial use under various trade and vernacular names. Generally speaking, Dark Red Meranti is sourced from Peninsular Malaya and Indonesia, Dark Red Seraya from Sabah and Sarawak, and Dark Red Lauan from the Philippines. This latter source was the main supplier a number of years ago, but large commercial quantities have been exhausted and lumber is now rarely exported. There is also a drive to add value in the country of origin, so less and less lumber is available as time goes by. The trees are tall, sometimes well over 60m (200ft) with clear, straight boles probably around 1m (3ft) in diameter. Timber cut from these boles is typically of dimension sizes; each thickness has boards produced in fixed widths. Also fairly long, probably starting at 2m (7ft) and typically going up to 5m (17ft) or sometimes over 6m (20ft) long. Greater lengths are possible, but become very difficult to handle. Subject to local variations, the sapwood tends to be slightly paler and sometimes difficult to define from the heartwood. The latter is normally a deepish red-brown colour. Some mineral deposits are also common and can be visible on the surfaces, showing up as a white or light grey streaking following the grain pattern. The grain is interlocked, giving a striped effect when cut on the quarter and is fairly coarse and open in appearance. Lumber can be affected by wood borers and pinhole or shothole evidence is often apparent.

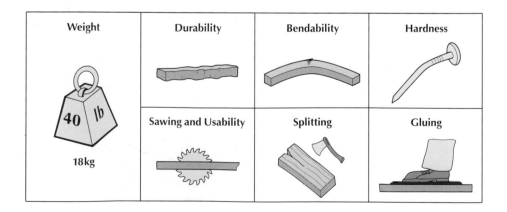

Weight	Durability	Bendability	Hardness
40 lb / 18kg			
	Sawing and Usability	Splitting	Gluing

Average dry weight of the Dark Red Shoreas is about 18kg (40lb), per cubic foot. It dries reasonably well but thicker material tends to check, split and distort although not in significant amounts. Shrinkage is minimal. It is not quite as strong as Beech. The heartwood is classified as moderately durable; treatment is recommended for external use.

Most Dark Red Shoreas will usually saw and plane successfully without too many problems. It nails reasonably well, but pre-drill near the end grain or when screwing. It glues and stains well. The surface will need some attention before a good polish is achieved.

This group of timbers can be classified as utility woods. They are usually employed for local construction work, joinery, shopfitting and similar applications. Often peeled plywood is available for a variety of uses.

Source:

South-east Asia.

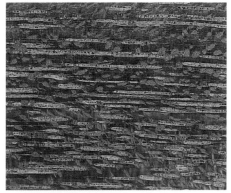

Radial

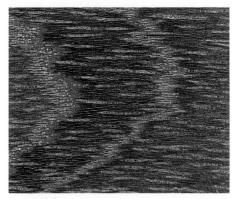

Tangential

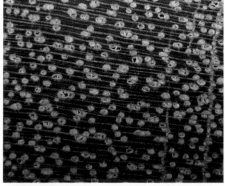

Cross Section

SHOREAS, LIGHT RED

Shorea spp.

Most commonly known as: Light Red Meranti, Lauan and Seraya.

The comments about sources and countries of origin applied to Dark Red Shoreas equally apply to this group. The trees associated with Light Red Shoreas tend not to be quite so tall probably only reaching a maximum of around 60m (200ft). Clear bole sizes are similar and therefore the lumber produced follows the same lines. Again, it is most typical to find dimension stocks prepared from these boles. Sometimes they will be cut into 'scantling' sizes, batched in thicknesses of 1.9cm (0.75in) thick upwards to fixed widths. When bought like this, wastage can be reduced and production maximized. The sapwood of the Light Red Shoreas is generally just discernible. The heartwood is a light reddish brown or pinkish brown colour. This colour does contrast and differ from its Dark Red cousin. Mineral deposits are also present, but they do not stand out as much against the lighter coloured wood. The grain is interlocked and when planed has a lustrous look and feel. It has a rather coarse but even texture that can be slightly wavy. When cut on the quarter a mild stripe is apparent. The wood is susceptible to attack by both pinhole and shothole borer. Occasionally, some slight blue staining can be found in association with the pin worm attack.

Light in weight, when dry it averages 15kg (33lb) per cubic foot. Not a strong wood, it has a rating some way below that of Beech. Some brittleheart or cross shaking can often be included. It dries rapidly with only some slight tendency to cup. Thicker

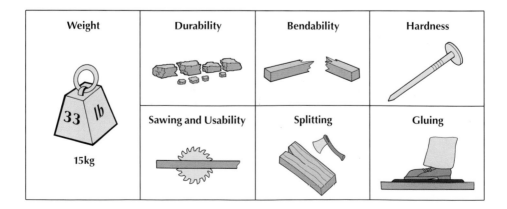

Weight	Durability	Bendability	Hardness
33 lb / 15kg			
	Sawing and Usability	Splitting	Gluing

material may surface check if the kilning process is not monitored correctly. There are higher levels of shrinkage associated with the Light Red than with the Dark Red. Although there may be some local variation, the heartwood is generally considered to be perishable.

Slightly softer than the Dark Red, this group works and machines in much the same way.

A lightweight timber, it is often utilized as a material from which to produce mouldings and trims. It is not particularly attractive and does not often get used for show wood furniture. Easily peeled, it is found as a core in plywood. It is another useful utility timber for internal use.

Radial

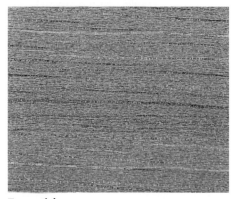

Tangential

Source:

South-east Asia.

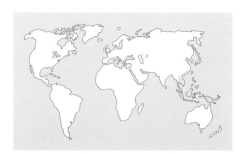

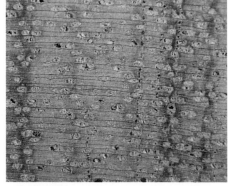

Cross Section

SYCAMORE

Acer pseudoplatnus

Also known as: Great Maple, Harewood, Plane, Scotch Plane, Sycamore Plane or Weathered Sycamore.

The often applied vernacular and trade names for Sycamore are confusing and in particular should not be used for European Plane, which is sometimes called Sycamore in Scotland. It should not be mistaken for American Sycamore (*Platanus occidentalis*), Sycamore is a closer relation to the North American Maple.

Sycamore grows throughout Europe; it is a fairly tall tree at around 30m (100ft) when mature. If grown in forest conditions it may produce clear, straight boles up to 15m (50ft) long; the majority will branch short of this. The wood is white with a slight creamy tinge with no clear demarcation between heart and sapwood. Diseased wood may have darker streaks or patches distributed throughout the boards. When quarter-sawn, a beautiful curly or fiddleback figure can be appear. The wood is prized for these figures, especially if it remains white in colour. The grain texture may be wavy or straight, and is normally fine and even. It can be chemically treated, which turns the wood a soft grey that is then marketed as 'Harewood'.

The dry wood will average 17kg (37lb) per cubic foot. It is not strong and is rated weaker than Beech. Straight, clear timber without any defects will bend easily. Sycamore is susceptible to staining, therefore it is crucial that some surface drying takes place immediately after conversion. A common trade practice is to up-end the boards and end dry them with small spaces between, not using

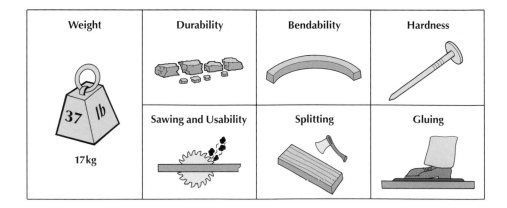

stickers at all. Once the surface has dried it should retain its colour unless it is placed in high moisture conditions. Sycamore performs reasonably well when kiln drying, although temperatures need to be mild to avoid any darkening. 'Weathered' Sycamore refers to timber that has been air dried for a long time and has gone a light brown colour.

Sawing and planing Sycamore is reasonably good, but some reduction of cutting angles may be required when planing to stop any tearing out of the grain. It turns, glues and nails well with only some pre-drilling required for screwing or when near the end grain. It takes stains well and a fine polished finish can be achieved.

This is a lovely timber to work with. It is an excellent wood from which to produce turned items and figured material is also prized for panelling.

Radial

Tangential

Source:

Throughout Europe.

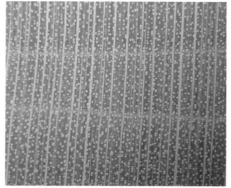

Cross Section

TEAK

Tectona grandis

Purists will tell you there is only one true Teak – however, traders may try to promote other timbers to emulate it! Teak is now grown in plantations in many tropical countries, but occurs naturally from India, Myanmar (Burma) through to the Indonesian island of Java, where it was introduced many years ago. The Teak tree grows up to a maximum of about 46m (150ft) tall. Depending upon local growing conditions, clear boles can be over 23m (75ft) long. More typically, they are around 15m (50ft) from Java and less when from Myanmar; bole diameters average around 1.2m (4ft). The trees are often 'girdled', in which the bark is cut away all around the trunk while it is still standing and is then left to dry for a number of years. This practice aids the eventual drying process. The sapwood is normally a narrow band of creamy coloured wood that is clearly visible. The heartwood has a greenish tinge when first cut, but this gradually darkens to a golden brown upon exposure. Some dark streaks can be visible depending upon place of origin; Teak from Myanmar tends to be more uniform in colour. There are natural oils present that contribute to Teak's reputation for durability. These oils provide a distinct musky smell that can be slightly overpowering when initially cut; it tones down after a while or goes off entirely if left completely exposed. The grain is generally straight although it can be wavy on occasion. A ring-porous timber, growth ring patterns are usually visible.

Teak's strength to weight ratio is excellent, rated on a par with Beech. An average dry weight is around 18kg (40lb) per cubic foot. Drying can be

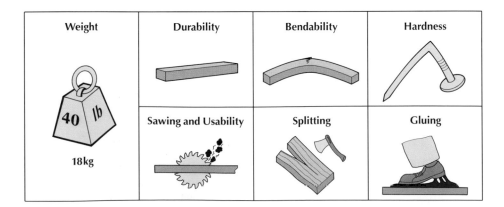

Weight	Durability	Bendability	Hardness
40 lb 18kg			
	Sawing and Usability	Splitting	Gluing

slow, with some pockets of moisture remaining and some end splitting. Colour variations may occur when kiln drying, but these tend to even out over time afterwards. Teak is naturally very durable.

This is an abrasive material, therefore blunting of saws and cutter blades takes place fairly rapidly. Keep tools sharp to ensure that satisfactory finishes are achieved. Pre-drill for all jointing operations. The natural oils can cause problems with some glues, so take care. Unlikely to need staining, Teak can be polished to a fine finish.

The list of end uses for Teak is long! Most commercial production goes into internal and external furniture, joinery and cabinet-making. Its properties also lead to a use as decking for yachts and boats, and as a flooring material.

Radial

Tangential

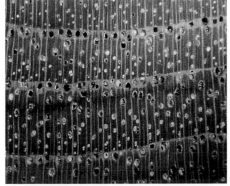
Cross Section

Source:

India, Myanmar and South-east Asia.

TULIPWOOD, AMERICAN

Liriodendron tulipifera

Also known as: American Whitewood, Canary Wood, Tulip Poplar, Tuliptree, or Yellow Poplar.

There is also an Australian wood, *Harpullia pendula*, sometimes called Tulipwood or Tulip Lancewood, that should not be confused with this species.

Commonly found in parklands, most commercial supplies of American Tulipwood come from the south and south-eastern USA. Occasionally timber from *Magnolia acuminata*, the Cucumbertree, is included in commercial parcels of lumber because of its close similarity to American Tulipwood. In favourable conditions, the tree grows up to 46m (150ft) with clear boles that have diameters of 1.8m (6ft) or more. When converted, wide planks are not uncommon. The sapwood is white in colour, as can be the heartwood of young, vigorously grown trees; older heartwood tends to be a light brown colour. Most often, the heartwood is a very pale yellow with a slight greenish hue. Dark streaks are also common and can be purple, green, blue or black; they are a distinguishing feature of this timber. The grain is usually straight, finely and evenly textured.

It is a lightweight wood, ranging between about 13 to 15kg (29 to 33lb) per cubic foot when dry. It is not strong, but can be quite stiff with a tendency to be brittle; it is therefore not a good bending material. It dries fairly easily, although there can be some noticeable shrinkage; once dry, it is considered to be a stable wood. American Tulipwood is not considered durable and should be treated with preservatives for any external applications.

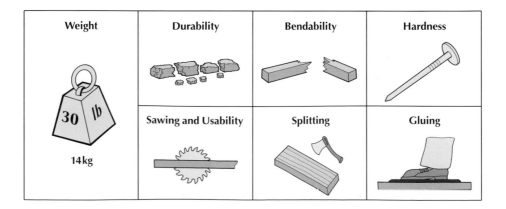

Weight	Durability	Bendability	Hardness
30 lb 14kg			
	Sawing and Usability	Splitting	Gluing

This wood is soft and easy to saw unless there has been some problem when drying that may lead to binding or burning. It planes and turns well, producing a nice smooth finish. Its stability and softness also make it ideal as a carving medium. It has good resistance to splitting, but it is still advisable to pre-drill when screwing. It takes glue well and can be stained uniformly. Although soft, a reasonable polished finish can be achieved if the timber is sealed correctly.

American Tulipwood is used extensively at source for a wide variety of uses. It has been utilized for internal joinery, panelling and carcass work. Some heavier examples can be used for furniture, but most would be for stuff over, drawer sides and similar applications. Poorer quality material is used up for pallet and packing cases.

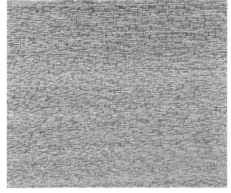

Radial

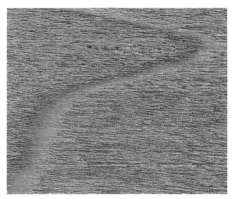

Tangential

Source:

Eastern North America.

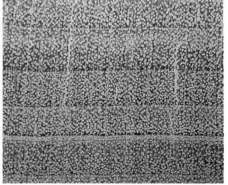

Cross Section

UTILE

Entandrophragma utile

Also known as: Assie, Kalungi, Mufumbi, Mvovo, Sipo, or Tshimaie.

Found throughout West and Central Africa, most of the commercial supplies of Utile come from West African countries. This is another member of the *Meliaceae* family and is related to Mahogany. Its closest cousin is Sapele, with which it is often confused or mistaken. The Utile tree is fairly tall at around 43m (140ft); it is not quite as tall as Sapele but is fatter. Clear boles of 21 to 24m (70 to 80ft) are not uncommon, with diameters of over 2.4m (8ft). The wood from these trunks is normally cut into random lengths and widths. The sapwood is a light brown colour and clearly distinct from the heartwood, which is a uniform reddish brown that tends to slightly darker than Sapele. The grain is open, irregular and interlocked, producing some striped effect although this is not pronounced. When freshly sawn, the timber does not have a particularly strong odour.

Utile's dry weight is around 19kg (42lb) per cubic foot on average. It is of equal strength to Beech and is fairly stiff; it does not bend well. It dries better than Sapele with less downgrade so long as there is not too much refractory grain present. Shrinkage is not excessive and once dry the wood is fairly stable. Utile is rated as a durable wood and resists treatment with preservatives rigorously!

When sawn, Utile is usually fairly mild to cut with only some occasional distortion off the saw. Planing is reasonable, but a reduction in cutting angle may help to avoid too much tearing out of the surface. When boring, some charring can take place

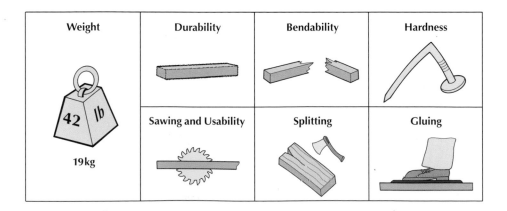

Weight	Durability	Bendability	Hardness
42 lb / 19kg	Sawing and Usability	Splitting	Gluing

if care is not taken. It can be turned and produces a nice finish. It nails reasonably well, but pre-drilling near the end grain is advisable and for screwing also. It glues well. Some surface preparation may be necessary prior to staining, if required, and polishing; a good finish can be achieved.

Utile can be used for any of the applications to which Mahogany or Sapele are applied. It is sought after for all internal and external joinery, particularly doors and windows. Even though it does not have such an attractive stripe as Sapele, it is still often used for veneers. It is a popular timber for show wood in upholstered furniture; solid wood furniture is also made from it. On occasion it has been used as a boat-building trim.

Radial

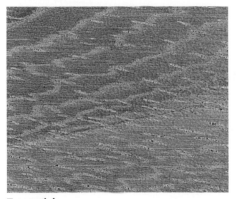

Tangential

Source:

Predominantly West Africa.

Cross Section

WALNUT, AFRICAN

Lovoa trichilioides

Also known by its country of origin for example Benin, Nigerian, and so on, and as Alonawood, Apopo, Bibolo, Bombolu, Congowood, Dibetou, Dilolo Fiote, Eyan, Lovoawood, Noyer d'Afrique, Noyer de Gabon, Sida or Tigerwood.

African Walnut can be found in West and Central Africa; the species is also found on the east coast but the wood is somewhat coarser in appearance. Commercial supplies have predominantly been sourced from West Africa. Looking somewhat like a Walnut tree, it grows up to about 40m (130ft) tall with bole diameters up to 1.2m (4ft). Clear boles are usually 18m (60ft) plus. Lumber is usually cut in random lengths and widths. A buff or dirty brown sapwood is clearly distinct from the golden brown heartwood. This latter will have darker streaks that give the wood a similar look to the true Walnuts. The moderately fine and even grain is shallowly interlocked, and stripes are produced on quarter-sawn faces. Occasionally, large holes are found in the heartwood upon conversion. Although called 'snake holes', it is most likely they will have been caused by a large beetle's lava. It is a matter of opinion why this wood is called Walnut at all, being more closely related to Mahogany. The name may originally have arisen simply because the tree or the wood bore some slight resemblance.

When dry, this timber weighs on average of 16kg (35lb) per cubic foot. It will dry fairly rapidly without too much shrinkage or distortion. It is not strong and can have a tendency to be slightly brittle. The heartwood is rated as moderately durable and does not take kindly to preservative treatment.

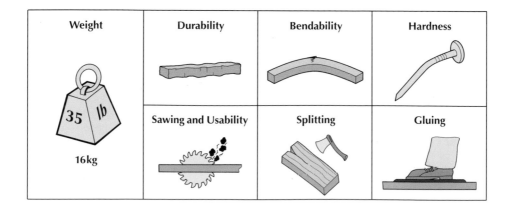

Weight	Durability	Bendability	Hardness
35 lb 16kg	Sawing and Usability	Splitting	Gluing

The slightly interlocked grain may affect the sawing and finish. To avoid too much tearing out of the planed surface some reduction of cutting angle helps. Although little blunting takes place, make sure all tools are sharp. The wood nails reasonably well, but pre-drill near the end grain and for screwing. It takes glues and stains well with little surface filling required before a satisfactory polish can be attained.

As one would expect, African Walnut has been used in similar applications to the true Walnuts, often disguised and sold as the real thing! In its own right it is a useful timber for furniture and cabinet work, panelling and mouldings. It should not be ignored because of its similarity to true Walnut, nor confused with it.

Radial

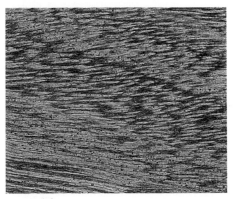
Tangential

Source:

Predominantly West Africa.

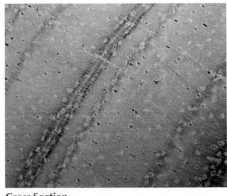
Cross Section

171

WALNUT, BLACK

Juglans nigra

Also known as: American Black or American Walnut – and sometimes as Gumwood.

Found naturally in central and eastern parts of North America, this Walnut is grown in other parts commercially for its nut crop and also as a decorative tree. It may reach up to 46m (150ft) high with bole diameters of more than 1.5m (5ft), if allowed to grow to full maturity. It gets its name from the distinctively patterned dark grey or black bark; the wood is also darker than its European cousin. When freshly cut, Black Walnut has a distinct and curious odour. This goes off after a while, although it can still be noticed when boards are separated from each other; it also has a mild taste! Together these lead to the old name of 'Gumwood'. A narrow, pale sapwood is clearly distinguishable from the rich chocolate brown of the heartwood. The colour is usually fairly uniform, despite the presence of some darker streaks. The grain is usually straight, although in older trees it may be curly or wavy, and the texture can be coarse. Boards containing the wavy and curly grained boards are sought after for veneering.

Black Walnut weighs, when dry, around 17 to 19kg (37 to 42lb) per cubic foot, 18kg (40lb) on average. It feels hard and heavy to the touch, with a strength value comparable to Beech and good resistance to shock. Drying can be difficult, depending upon the grain structure, with some occurrence of honeycombing. Shrinkage is not a major factor and the wood is considered to be stable after drying if kept in consistent atmospheric conditions. The heartwood is rated as being very durable and should not be wasted in

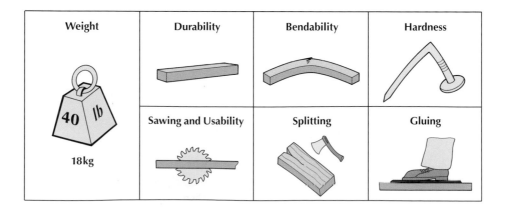

Weight	Durability	Bendability	Hardness
40 lb / 18kg			
	Sawing and Usability	Splitting	Gluing

external uses where preservatives may be required.

A fairly mild timber to handle, it cuts and saws well. Keep blades and knives sharp, especially if any of the grain is not consistently straight. This wood turns well and will produce a good surface finish. If nailing or screwing, it is best to pre-drill; it takes glue well. If stains are to be applied it will take them evenly. Some surface preparation may be necessary before polishing it to a fine finish.

Black Walnut is a beautiful wood that is sought after for high-class furniture and cabinet work in both the solid and veneer form. This latter is especially attractive if cut from burrs or crotches. It is also used as a traditional wood for gun stocks where its shock-resistant properties excel.

Radial

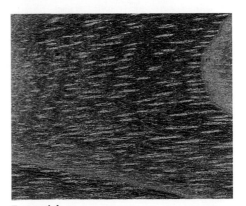

Tangential

Source:

North America.

Cross Section

WALNUT, EUROPEAN
Juglans regia

It can be prefixed with country of source, for example English, French and so on, and is also known as Circassian Walnut.

Probably originating in and around Turkey, this Walnut was introduced by settlers into many countries, spreading far and wide over the centuries. Cultivated for its nut, a delicious and nutritional food, and for the extraction of its oils. The green husk of the nuts and leaves produces a strong scent and will stain hands brown. The tree is slightly smaller than its American cousin, probably reaching only 30m (100ft) at maturity. Diameter of boles can be up to 1m (3ft) or so but any wider is fairly rare. The bark is a distinct grey colour that is much lighter than Black Walnut; new growth is smooth, old cubed and deeply furrowed. A thin, pale, clearly defined sapwood leads to a warm grey-brown heartwood that has a hint of purple or chocolate brown. The texture of the wood is fairly coarse and feels cool to the touch. When grown in stands the grain will be reasonably straight, but most park or field Walnut will have variable and interlocked grain, especially in older trees. Any examples with interesting growths will be prized for their possible veneer value.

Slightly lighter in dry weight to its cousin, European Walnut averages about 17kg (37lb) per cubic foot. It has a hard surface and good resistance to shock, with a strength comparable to that of Beech. It dries well but can honeycomb; some distortion may be apparent with wild-grained boards. A stable wood when dry, the heartwood is considered moderately durable, but it should not be wasted in external usc.

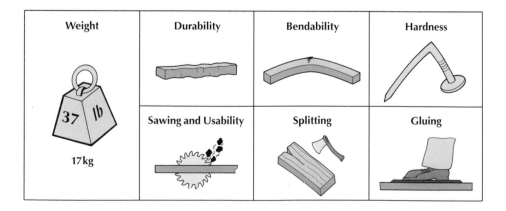

Weight	Durability	Bendability	Hardness
37 lb / 17kg			
	Sawing and Usability	Splitting	Gluing

The wood is reasonably good to work, except when an irregular grain is present, but all tools should be kept sharp. This is a nice wood to turn and can be finished well. If nailing or screwing, pre-drill first. It glues well and takes a stain fairly evenly. An exceptional polished surface can be achieved with effort.

Like all true Walnuts, this wood is sought after for use in quality furniture and cabinet work. Figured French Walnut is exceptionally prized for use in gun stocks where its beauty can shine through! Any appendage or slight defects in growth will produce some outstanding veneers; these are often seen in antique furniture. It is a fine wood to use for panelling and decorative mouldings. It can also be carved fairly easily and is often used as a medium for this.

Radial

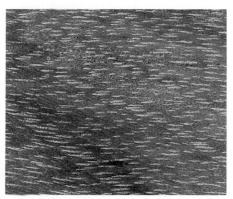

Tangential

Source:

Throughout Northern, Central and Southern Europe.

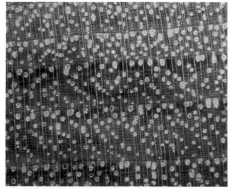

Cross Section

175

WENGE

Millettia laurentii

Also known as: Dikela or Palissandre du Congo.

This is a West and Central African tree that should not be confused with Panga Panga (*Millettia stuhlmannii*), which, although it has very similar properties, grows in East Africa. Not a large tree, Wenge probably only grows to a maximum of around 18m (60ft) in height. Straight boles are short, only about 5m (17ft) long on average and only about 60cm to 1m (2 to 3ft) wide. This is reflected in the short and narrow lumber that is produced. When freshly sawn, Wenge has a slightly musky smell that goes off after a time. The sapwood is a distinctive yellowish brown; the attractive heartwood is a deep, dark brown with alternating depths of colour that produce a narrow stripy effect. The grain is very open, coarse and usually irregular. It is susceptible to attack by pinhole beetle, the damage of which is difficult to spot on the nearly black background! The wood feels cold and hard to the touch, with a slight oily or greasy feel to it.

This is a heavy and hard wood weighing, when dry, anything from 25 to 30kg (55 to 66lb) per cubic foot, with an average of 18kg (40lb). It has exceptionally good resistance-to-wear properties. The wood is quite brittle and has a tendency to split fairly easily. It dries slowly and reasonably well, but any existing splits will possibly get worse during drying and there may be some tendency towards surface checking. Once dry, there is little further movement so long as it is kept and used in favourable conditions. The heartwood is considered to be durable and both it and the sapwood are highly resistant to treatment with preservatives.

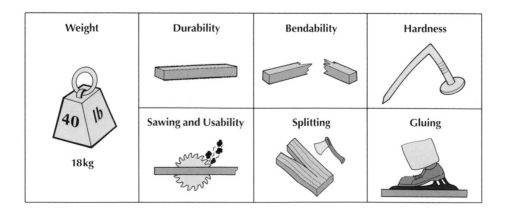

Weight	Durability	Bendability	Hardness
40 lb 18kg			
	Sawing and Usability	Splitting	Gluing

Wenge can be difficult to machine because of its hardness. When sawing, some binding and burning can take place; it will also chatter when planed if not held firmly. All tools should be kept sharp. Spelshing can be a problem when cross-cutting unless the back face is fully supported. This is a good timber to turn. It tends to split, so pre-drill when fixing. Gluing is satisfactory. It is hard enough to produce a good polished finish.

This is a wood that has been rather overlooked for decorative purposes. It can be used for trims and mouldings to contrast with other woods or turned for fancy goods in its own right. Its resistance-to-wear properties make it an excellent wood for flooring, where its striking appearance can be an additional attraction.

Radial

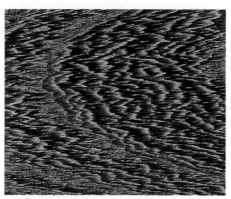

Tangential

Source:

West and Central Africa.

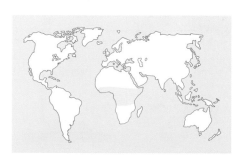

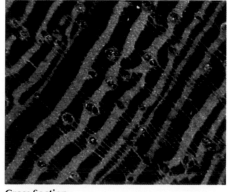

Cross Section

ZEBRANO

Microberlinia brazzavillensis

Also known as: Zingana and sometimes Zebrawood (the use of the name 'Zebrawood' is confusing because several other African timbers are known by this name).

Zebrano is found in West Africa, where limited commercial quantities are occasionally available. The tree is not tall, rarely attaining heights of over 20m (65ft) high. Boles are generally around 60cm to 1m (2 to 3ft) in diameter. After felling, the sapwood is sometimes adzed off, leaving only the decorative heartwood for commercial use. When lumber is available it will be found in short and narrow specifications. The sapwood is pale and clearly distinct from the heartwood that gets its name from the narrow, alternating bands of colour. The lighter colour can be a bright or creamy yellow with dark brown or black narrow strips. This stripy effect

is best seen when the wood is cut on the quarter, where it tends to be more evenly spaced. If flat sawn or rotary peeled the alternating colours are more random. The wood feels hard to the touch and some deposits of resin are often present. The grain is slightly interlocked, and is coarse and open in texture.

Zebrano is quite a heavy wood when dry. It weighs anything from 18 to 21kg (40 to 46lb), averaging 19kg (42lb), per cubic foot. It is noted for its shock resistance and is fairly strong, with ratings slightly better than Beech. This can be a difficult timber to dry with a tendency to surface check and end split. It is advisable to pre-dry on thin stickers before kilning to help avoid too much rejection. It will also tend to twist as well. It is a good idea to weigh down air or kiln drying stacks so as to

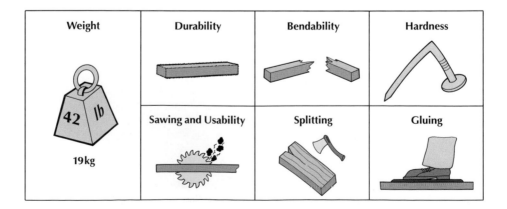

Weight	Durability	Bendability	Hardness
42 lb 19kg			
	Sawing and Usability	Splitting	Gluing

alleviate this. Little is known about Zebrano's durability, as it is too rare and expensive to be used for anything other than quality work, but it is likely to be durable at least.

This wood saws fairly easily but may be difficult to plane, so keep all tools sharp. Pre-drill when fixing because of the tendency to split. It glues well, is unlikely to be stained and can be polished to a fine finish, although some surface filling may be required.

Zebrano is rarely used as a solid wood, but can be so used for fine cabinet and furniture work. When used as a veneer, it is popular as a cross banding for inlay work or in larger sections for panelling.

Radial

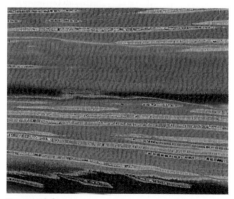

Tangential

Source:

West Africa.

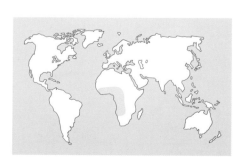

Cross Section

CALIFORNIAN REDWOOD

Sequoia sempervirens

Also known as: Coast Redwood, Redwood or Sequoia.

A large tree that can grow to 91m (300ft) or more; the Californian Redwood is not as big as its cousin, *Sequoiadendron giganteum*, which is noted for being one of the longest living and largest trees on earth! The Californian Redwood is only found on the west coast of America, growing in a strip near to the coast, and commercial production is mainly limited to California. Slightly tapered boles can be clear for up to 60m (100ft); at the buttressed end they may reach diameters of 3.6 to 4.5m (12 to 15ft). The tree has a very thick bark, up to 30cm (12in) or so. This bark has properties that are utilized in their own right for specialist products such as filters. Felling of Californian Redwood is controlled. Only a limited amount comes to the market each

year, therefore it attracts a premium. The heartwood is of a fairly straight-grained appearance, reddish or purple-brown in colour, with a thin band of yellowish sapwood; there is no evidence of resin. Slow growing, the growth rings are fairly conspicuous. The grain generally has a mild, even texture, but can be coarser depending upon local growing conditions.

A lightweight wood, this can work against it at times. On average it weighs, when dry, around 12kg (26lb) per cubic foot. It is soft and easily dented with a fingernail. It dries well, on a fairly high kiln schedule, with little distortion or downgrade except for some end splitting. Care with stickers is advisable when kilning thinner stock. Once dry, it is considered to be a fairly stable timber. Californian Redwood is not particularly strong and when

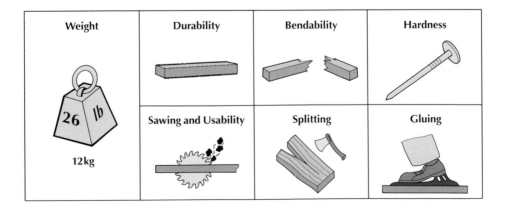

Weight	Durability	Bendability	Hardness
26 lb / 12kg	Sawing and Usability	Splitting	Gluing

compared with European Redwood will have values slightly below. Its heartwood is considered to be durable and it may resist preservative treatment.

This wood is quite easy to work, but all tools need to be kept sharp or the finished surface may appear woolly. Good extraction is required when planing to avoid any chip bruising. It does splinter and split, therefore pre-drill when fixing, watching that they do not pull out! It glues well, although some adhesives may mark the surface. It can be stained and polished but is easily dented.

Californian Redwood, although soft and light, makes the most of its durability for shingles and similar applications. It is a useful timber for joinery provided that is not exposed to too much wear.

Radial

Tangential

Source:

West coast of USA.

Cross Section

CEDAR

Cedrus spp.

Also known as: Atlantic or Atlas Cedar, Cedar of Lebanon and Deodar.

Because of a possible similarity in smell, the name 'Cedar' is often used in association with both hardwoods and softwoods; however, there are only three true Cedars. Atlantic or Atlas Cedar (*Cedrus atlantica*) is found in the North African countries of Algeria and Morocco, growing to a maximum height of around 46m (150ft). Cedar of Lebanon(*Cedrus libani*) comes from the Middle East and grows to a similar height. Deodar (*Cedrus deodara*) is a native of the northern Indian Himalayan regions and is the largest of the Cedars growing anything up to 60m (200ft). All of these Cedars can be found growing outside their indigenous sources as ornamental trees in parklands and gardens. The wood from all the true Cedars is very similar. With a pale, reasonably distinct sapwood, the heartwood is usually a light brown in colour. The wood has a strong, persistent odour that is easily recognized. Growth ring markings are apparent, with clear contrasts between the faster grown earlywood and the denser latewood. The grain is fairly straight with a medium to fine texture.

It is a medium-weight wood when dry, at around 16kg (35lb) per cubic foot. It dries reasonably quickly, but extra stickers and weighing down are recommended to avoid high levels of distortion. Although similar in strength values to European Redwood, it does tend to be brittle and the surface is of medium hardness. The heartwood is considered to be durable and resists preservative treatments.

The true Cedars saw and plane well to a good surface finish. Some

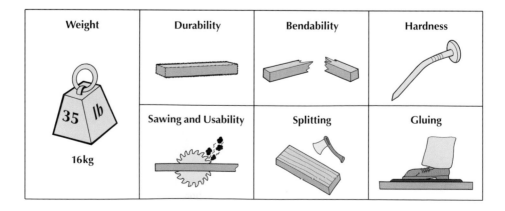

Weight	Durability	Bendability	Hardness
35 lb / 16kg			
	Sawing and Usability	Splitting	Gluing

care needs to be taken with knotty material to avoid chipping out, and spelshing can be a problem when cross-cutting; keep all tools sharp. It resists splitting and can be nailed easily, but pre-drilling is always advisable when screwing. It will glue well, take stains and can be polished to a fine finish.

In source countries, true Cedars are often used in demanding applications such as sleepers, bridge construction and house building as they have a reputation for longevity. When used for joinery and furniture they can be very decorative and occasionally veneers are available. Outside the countries of origin, trees grown in parks and gardens tend to be used for fencing, gates and other similar uses. It is suitable for use in garden furniture.

Radial

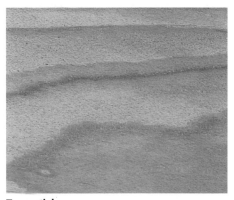

Tangential

Source:

North Africa, Middle East and Northern India.

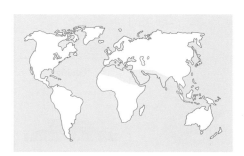

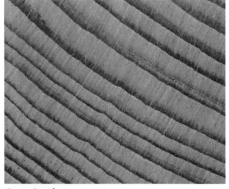

Cross Section

CEDAR, PENCIL

Juniperus spp.

Also known as: African, East African or Virginian Pencil Cedar and Eastern Red Cedar.

These are not true Cedars. The original Pencil Cedar (*Juniperus virginiana*) is found in eastern North America. This was extensively used until supplies started to become limited, then another source (*Juniperus procera*) was found and utilized from East Africa. Both trees have similar properties and the distinct smell and taste of pencils! The trees grow anything up to 30m (100ft) if allowed to mature, with diameters of up to 1.5m (5ft). The North American variety is slightly shorter and will tend to buttress at the bottom. The heartwood has a fine, even texture; initially, its colour will be a pink purplish red that darkens down upon exposure to a more even reddish brown. Growth rings are apparent, although more so in the North American wood. The sapwood is clearly distinct and is cream or dirty white.

Pencil Cedar is a medium-weight wood, with the African variety being slightly heavier than the American. On average, the dry weight will be around 16kg (35lb) per cubic foot. Both timbers are slightly stronger than European Redwood, and are hard and resistant to wear. The wood is not easy to dry, with significant amounts of surface checking, distortion and end splitting. It should be dried slowly and with care, especially the thicker material. Shrinkage is not great and once dried to its final moisture content it is considered to be a stable timber. The heartwoods of both varieties are rated as durable and are resistant to treatment with preservatives.

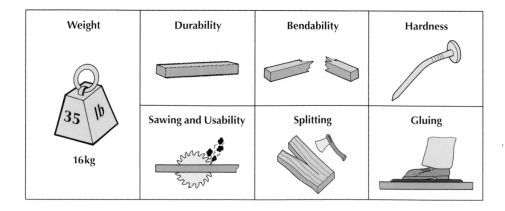

Sawing and planing Pencil Cedar can be accomplished fairly easily with sharp tools, although some tearing or breaking out may occur around knot holes. On the back faces of cross-cut boards and mortice holes, provide support to avoid small chips being knocked out because of the wood's brittle nature. All fixings will need pre-drilling to avoid splitting; it glues reasonably well. It takes stains readily and a good surface finish and polish can be achieved.

Pencil Cedar is used traditionally for production of high-class pencils, but also has uses for joinery, furniture and other products. It has been used to line fancy boxes and chests where its natural aroma can be appreciated. Mountain and shrubby forms are often used as flavourings and distilled for oils. It is also a good wood to whittle!

Radial

Tangential

Source:
Eastern North America and East Africa.

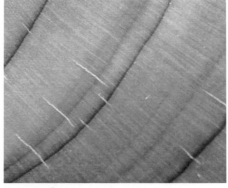

Cross Section

CEDAR, PORT ORFORD

Chamaecyparis lawsoniana

Also known as: Lawson's Cypress, Oregon Cedar and White Cedar.

Because it has a somewhat similar fragrance to the real thing, this is another tree that is called by the name Cedar. It grows naturally in the USA on the western seaboard, but can be found in other countries as an exotic park or garden tree. In these circumstances it is most often known as Lawson's Cypress. Port Orford Cedar can grow to heights of around 60m (200ft) when mature. Bole diameters up to 3.6m (12ft) above the swollen base are also possible. Large commercial quantities are in limited supply. When cut, the timber has a fine and even texture with a distinct spicy odour. The heartwood is difficult to distinguish from the sapwood. It has a pale pinkish yellow or light brown colour, typically without resin. Some small resin beads can occasionally be seen on sawn and planed surfaces. Growth rings are not always apparent as the change between the early and latewood is gradual.

This is a medium to lightweight timber when dry. It will average around 13kg (29lb) per cubic foot. It has some good strength to weight properties and is rated significantly stronger than European Redwood. It is fairly hard and has moderate resistance to wear; it does not like to be bent! Port Orford Cedar is reasonably easy to dry, with little degrade and only a moderate amount of shrinkage. The heartwood is considered highly resistant to decay and is therefore rated as durable; it is moderately resistant to treatment with preservatives.

This wood will work reasonably well so long as the grain is uniformly

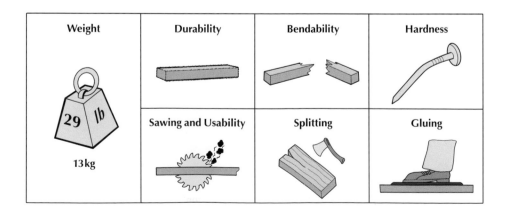

straight. Any examples with irregular grain will need more care to avoid tearing out, especially if knots are included. If resin is present it can gum up the sawteeth and planer knives; always keep cutting tools sharp. Although it nails reasonably well, pre-drill near the end grain and when screwing to avoid disappointment! It glues readily and should stain fairly evenly. Resins can cause problems and surfaces need sealing but a good finish can be achieved.

Port Orford Cedar has traditional uses in ship and boat-building, door and window construction, furniture, flooring, cabinet work and interior trims. It has also been used for sports equipment such as canoe paddles and archery parts. It is also used as a material from which to make matches.

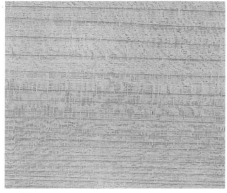

Radial

Tangential

Source:

West coast of USA.

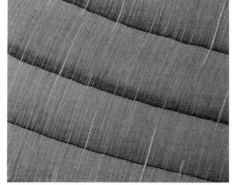

Cross Section

187

CEDAR, WESTERN RED

Thuja plicata

Also known as: British Columbian Red Cedar, Giant Arborvitae and Red Cedar.

Not a true Cedar, the Western Red Cedar may have some similar characteristics. The tree is found throughout North America, but most commercial supplies will come from west of the Rocky Mountains in Canada and north-west USA. Some plantations can be found in New Zealand and the UK, where it grows reasonably well. A tall tree, in its natural habitat, reaching heights of up to 76m (250ft); boles can be over 1.2m (4ft) in diameter. The whitish, narrow band of sapwood is clearly discernible. The heartwood is anything from a pale pink through to a reddish brown initially, but darkens to a more uniform colour on exposure. Often used in external applications, the surface of Western

Red Cedar will weather to a light silvery grey. When freshly cut, the timber has a distinct smell similar to the true Cedars. This odour can be persistent and is apparent when boards in piles or stacks are disturbed. The wood is non-resinous and usually has a straight grain of a fine to medium texture.

The lightest of the generally available softwoods when dry, Western Red Cedar will average around 10kg (22lb) per cubic foot. It dries rapidly and well in thinner sizes, but material over 5cm (2in) can tend to be troublesome; pockets of moisture will remain however long one tries to remove them! It is possible during extended drying periods that a certain amount of collapse may occur. This is a soft, slightly brittle wood that is significantly less strong than European Redwood; the surface can

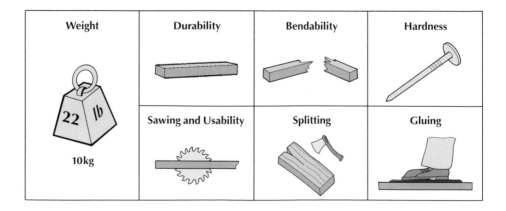

Weight	Durability	Bendability	Hardness
22 lb / 10kg			
	Sawing and Usability	Splitting	Gluing

easily be dented. The heartwood is considered durable and is resistant to preservative treatment.

Provided that saws and cutters are kept sharp, the finished product from Western Red Cedar is good. It does have some resistance to splitting but pre-drilling near ends when nailing and screwing is advisable. The wood contains chemicals that will react with ferrous metals and cause a stain; use galvanized copper or other treated fixings to avoid this. It glues, stains and polishes well but can be easily marked.

This wood has been used traditionally for shingles and other exterior claddings. It is not strong and should not be used for structural purposes. It is popular with conservatory and greenhouse manufacturers because it is easy to work and durable; it is also the first choice of wood when making beehives!

Source:

North America.

Radial

Tangential

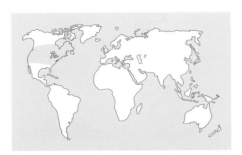

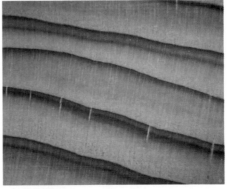

Cross Section

CYPRESS

Cupressus spp.

Also known by its country of origin, for example Mexican, or from a region, for example Monterey.

From the warm, temperate regions of the world, Cypress has been of only local significance until recent times. Various members of the *Cupressus* species are now grown as a plantation tree in climates slightly more conducive to rapid growth. It does well on poor soils and the main areas of cultivation are to be found in Australia, New Zealand and South and East Africa. Within twenty-five to thirty years a tree of commercial size can be grown and cropped from these plantations. The Cypress tree naturally grows to heights of up to 21m (70ft) with bole diameters up to 1m (3ft). *Cupressus lusitanica*, grown in Mexico and Guatemala, tends to be taller at around 30m (100ft). Cypress heartwood is generally a pale yellow or pinkish brown with a lighter coloured sapwood. Resin cells are occasionally visible as streaks or darker marks but these do not cause a problem. The grain is fairly straight, although plantation-grown material can have a preponderance of knots. The texture is fine and even; the growth rings are not distinctly visible. When freshly sawn, the wood gives off a smell similar to that of Western Red Cedar but this soon disappears.

Cypress is not a heavy wood when dry. It will average about 13kg (29lb) per cubic foot. It dries fairly rapidly without a significant amount of downgrade. It can be brittle and is not a strong wood with a value rated some way below that of European Redwood; its surface is easily dented. The wood is considered durable and suitable for use in contact with the ground.

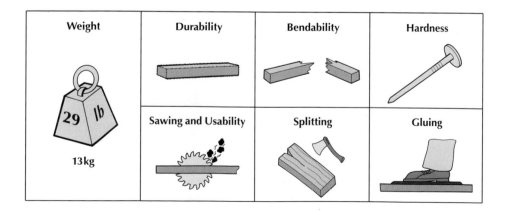

Weight	Durability	Bendability	Hardness
29 lb / 13kg	Sawing and Usability	Splitting	Gluing

Sharp saws and planer knives will produce satisfactory finishes on Cypress. A little care will need to be exercised when cross-cutting; there is a tendency for the wood to spelch out on the back side. Some plantation-grown material with uneven grain or a large amount of knots will also need some care to avoid tearing out. The timber nails well and should only need pilot holes when screwing; it glues well. It will take stains, but be a bit cautious, in order to prevent patches occurring; it can be polished to a reasonable finish.

Cypress is used in countries of origin for constructional work, for posts and poles, and so on. Some selected timber is suitable for joinery and external applications.

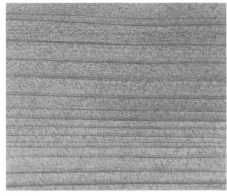

Radial

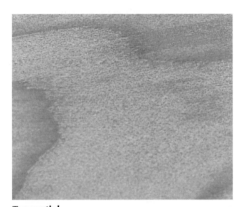

Tangential

Source: Southern USA, Central America and South America, East and South Africa, Australia and New Zealand.

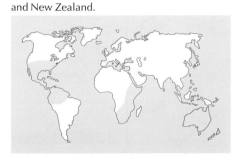

Cross Section

191

DOUGLAS FIR

Pseudotsuga menziesii (formerly *P. taxifolia* and *P. douglasii*)

Also known as: Red and Yellow Fir, British Columbian or Columbian Pine and Oregon Pine, and Douglas Spruce.

Although often called by a variety of misleading names, Douglas Fir is a distinct and separate species. Probably one of the best known of the softwoods, this tree can grow to heights of over 91m (300ft) in old growth forests. More typically, it will reach between 46 to 60m (150 to 200ft), of which two-thirds will be a clear, slightly tapering bole up to 2.4m (8ft) in diameter. From these trees significant amounts of clear grade lumber can be produced. Although now grown throughout the world, Douglas Fir is found mainly in North America, with most commercial supplies coming from British Columbia in Canada and the Oregon and Washington states in the USA.

The sapwood is not normally much wider than about 5cm (2in) and is slightly lighter in colour than the heartwood. The heartwood has clearly defined early and latewood growth rings, producing a distinct wavy pattern on tangential faces and when rotary cut; quarter-sawn material is usually very straight grained. The wood ranges from a pale to dark brown over these growth rings, but can have a reddish hue that leads to its name of Red Fir. Some resin ducts and pockets are present.

This is a medium-weight softwood. When dry, it will average between about 16kg (35lb) per cubic foot. It is rated as much stronger than European Redwood and is more resistant to bending. It has good resistance to wear when cut on the quarter; its values are nearly up to those of Pitch Pine. Thin material will

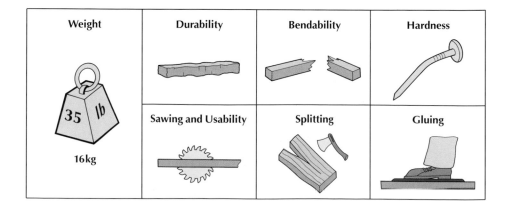

Weight	Durability	Bendability	Hardness
35 lb **16kg**			
	Sawing and Usability	Splitting	Gluing

dry fairly rapidly without too much degrade. Thicker material needs more care to avoid surface checking and end splitting; quarter-sawn stuff is much easier. Shrinkage is not significant, although the wood can pick up and give off moisture easily; this leads to movement *in situ*.

Douglas Fir works reasonably well with sharp tools but has a tendency to split, therefore pre-drill for all fixings; it glues well. Staining can be patchy depending upon the early and latewood growth. It can be polished to a satisfactory finish.

This is a well known timber for top quality joinery, and construction and structural work. It can be peeled for use in exterior grade plywood. Quarter-sawn, 'vertical grain', material can be used for flooring and is sought after for frame and ladder work.

Radial

Tangential

Source:

North America.

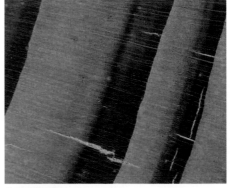

Cross Section

FIR, SILVER

Abies alba

Also known as: European Silver Pine or Whitewood.

Only the *Abies* species produces true Firs, although others are sometimes referred to by this name. Silver Fir is a member of this species and is found growing naturally throughout Northern, Central and Southern Europe. It is often included within commercial parcels of timber in Europe marketed, along with Spruce, as 'Whitewood'. The tree grows under natural, favourable conditions to a maximum height of around 46m (150ft) with 1.8m (6ft) diameters; more generally they are somewhat smaller than this. When sawn, there is no discernible difference between the sap and heartwood. The wood is coloured a pale creamy white, sometimes with a yellowish tinge. It is a rather bland wood, without clearly defined early and latewood growth patterns. The grain is usually straight and the texture fine and mild with little lustre.

A lightweight wood when dry, it averages around 13kg (29lb) per cubic foot. It has a fairly hard surface when dry and matches European Redwood in strength values; the wood is fairly brittle. It dries reasonably rapidly without too much distortion. Some checking and splitting will occur and knots can loosen and drop out; shrinkage is noticeable but not significant. Silver Fir is not classified as durable and is generally restricted to internal use; if used externally it should be treated with preservatives.

With sharp saws and planing knives a good finish can be achieved. Silver Fir is often moulded on planer moulding machines, through which it can be fed at fairly high speeds, making it an economic timber from

Weight	Durability	Bendability	Hardness
29 lb / 13kg			
	Sawing and Usability	Splitting	Gluing

which to produce mouldings. When nailed, it will not split except near the end grain; a pilot hole may be needed when drilling. It glues well and will stain fairly evenly. Not often polished, it is a good material to paint; knots will need to be treated with 'knotting'.

Silver Fir is a utility timber that is generally found on the market mixed in with other species and sold as Whitewood. It can be used as a carcassing material and is employed extensively for internal joinery and mouldings with a painted finish. It can be used for other carpentry work and in non-show wood positions in furniture. It is not durable and if used for windows and doors will have a limited lifespan. It is often used as a base pulping material for the production of paper and boards.

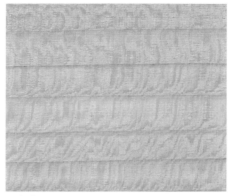

Radial

Tangential

Source:

Throughout Europe.

Cross Section

HEMLOCK, WESTERN

Tsuga heterophylla

Also known as: Grey Fir, British Columbian, Pacific or West Coast Hemlock, Western Hemlock-Fir, Hemlock-Spruce or Alaska Pine

North America is the only significant source of commercial Hemlock and, as some of its other names infer, it is to be found in the central and western regions of both Canada and the USA. Like a lot of softwoods, it has a mass of other local names that will always add to the confusion of identification. Western Hemlock has been introduced into other countries and tends to be grown in mixed stands or plantations. The tree, in its natural habitat, grows up to around 60m (200ft) in height and may measure up to 2.4m (8ft) in diameter. When cut, there is a clear distinction between the sap and heartwood. The former is a paler version of the latter, which is a yellowish brown with a darker, possibly slightly reddish or purplish brown latewood. Growth rings are easily apparent but not bold; the grain is usually straight. Dark streaks can often be seen; these are caused by an attack of Hemlock Bark maggots, but do not affect the strength properties of this wood. In addition to this, some thin darker lines can also be seen on occasions. These are traumatic resin ducts; they will not leach out and bleed through a finished surface.

When dried, Western Hemlock will have an average weight of around 14kg (30lb) per cubic foot. Its strength values are rated as similar to those of European Redwoods; its surface is reasonably resistant to denting. This wood dries fairly slowly and care needs to be exercised, especially with thicker material, to avoid surface checking. Shrinkage is

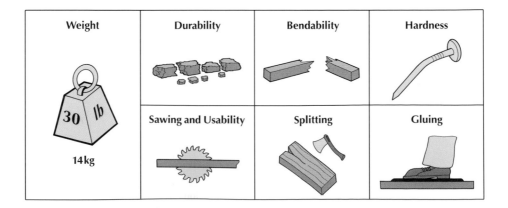

Weight	Durability	Bendability	Hardness
30 lb / 14kg			
	Sawing and Usability	Splitting	Gluing

noticeable but not significant; thicker material can produce patchy drying results. The heartwood is not classified as durable and resists treatment with preservatives.

Western Hemlock works reasonably well with little blunting, but keep all tools sharp to produce the best finishes. Some breaking out and spelching may occur on back edges if the timber is not fully supported when cut. It nails reasonably well but not near any end grain! Pre-drill for this and when screwing. It glues, stains and paints well.

This timber is used for internal mouldings and trims such as spindles for stairs. At source, large quantities are used for all sorts of constructional work including case-making, joists, studding and plywood. Some of the better quality material is used for joinery and ladder parts.

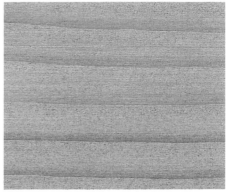

Radial

Tangential

Source:

Western North America.

Cross Section

197

LARCH, EUROPEAN
Larix decidua

This is one of a very limited number of softwoods that has a deciduous habit, shedding its needles over winter. It is naturally found throughout Central and Southern Europe, but it is now grown in plantations all over the UK, Continental Europe and elsewhere. The tree grows to a maximum height of around 46m (150ft) with a bole diameter of up to 1.2m (4ft). If grown without care, the boles will not be clear of branches for any length and in these instances the timber may tend to be very knotty. When not grown in tight stands, European Larch tends to develop a spiral grain that leads to problems during sawing. It is not often mixed or confused with other softwoods. This is because it is slightly heavier and has very wide, distinctive growth rings, which are formed by the early and latewood

configurations. The later heartwood is a dark reddish to brick brown colour with a contrasting lighter earlywood. The sapwood is clearly distinct and of a paler colour. Additionally, there can often be seen small, darker resin flecks on the planed surfaces of the timber; the grain is usually straight.

European Larch is one of the heavier softwoods when dry. It can range from 14 to 18kg (30 to 40lb) per cubic foot. It will dry fairly rapidly but has a tendency to distort, especially if spiral grain material is present; shrinkage is significant. This is most noticeable around knots, to the extent that they may drop out! It is probably one of the harder softwoods with strength values just in excess of European Redwoods. Although considered to be durable, the heartwood is in fact rated as only moderately so; it is extremely resistant

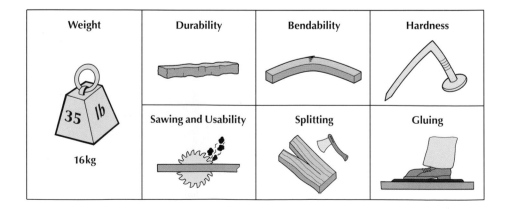

Weight	Durability	Bendability	Hardness
35 lb 16kg			
	Sawing and Usability	Splitting	Gluing

to treatment with preservatives.

This timber saws and planes fairly readily, although some gumming up from the resins will occur. Poorly dried material will produce great quantities of waste and any material with spiral grain will be very difficult to saw and keep flat. It does have a tendency to split, therefore pre-drill before nailing and screwing; ferrous metals will mark the wood. European Larch can be glued, stained and painted but some patchy results may occur.

Top quality, clear and straight timber is sought after as a 'boat skin' material, but most of the wood from this tree will be used for general agricultural purposes, telegraph poles, fencing and constructional uses. It also plays an important part in the production of turpentine and tannins.

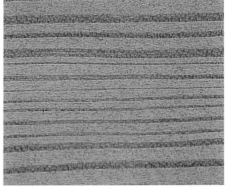

Radial

Tangential

Source:

Throughout Europe.

Cross Section

LARCH, WESTERN

Larix occidentalis

Found in western North America this Larch, like its European cousin, is one of the heavier softwoods to be commercially available. The Western Larch grows in British Columbia and in northern Oregon, into Montana and Washington states in the USA. Taller than the European Larch, when mature it can reach heights of up to one 55m (180ft) with similar bole diameters. It has the characteristic wide growth rings created from the late and earlywood growth patterns. The sapwood is a paler colour compared to the heartwood, which varies from a yellowish brown to a darker reddish brown. The wood is very resinous and slightly scented. With trees grown in more open or parkland settings there can be a preponderance of small knots, all of which will probably be darkly coloured. Most of these will be tight, live knots.

All Larches are considered to be generally heavier than other softwoods, and Western Larch is no exception. When dry, it will weigh around 17 to 19kg (37 to 42lb), 18kg (40lb) on average, per cubic foot. It dries fairly rapidly, but with a tendency to warp, split and have ring shakes. Shrinkage is significant and any dead knots are likely to drop out after drying. Western Larch's strength values are similar to European Larch; it is rated stronger than European Redwood. It is fairly stiff and splits easily. Although often used in external applications, it is classified as only moderately durable; it is very resistant to preservative treatment.

Western Larch has many of the same working properties that apply to European Larch. It will saw and plane

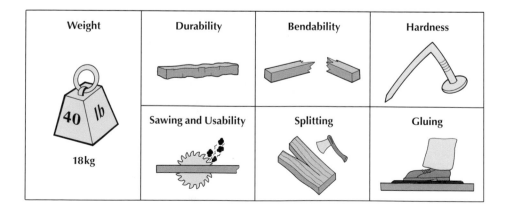

Weight	Durability	Bendability	Hardness
40 lb **18kg**			
	Sawing and Usability	Splitting	Gluing

reasonably well, but some blunting may take place. If particularly resinous boards are being machined, regular cleaning will need to take place. It will split easily, so pre-drill for all fixing operations. It glues reasonably well in most cases but the resin content will have a bearing upon success! When painting, the wood must be dry and it will need coating with a proprietary sealer to ensure that no resins come through.

Western Larch is used for many external applications, such as rough boards, poles, posts and fencing, cross ties, railway sleepers and mine timbers. It has similar properties to Douglas Fir and can often be found mixed in the same parcels. Some of the better quality material is suitable for joinery, door manufacture and flooring.

Source:

North America.

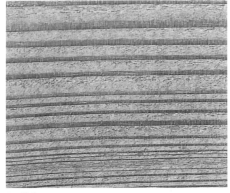

Radial

Tangential

Cross Section

201

PARANA PINE

Araucaria angustifolia

Also known as: Brazilian Pine, Curi-y, Kuviy, Pinho, Pinho de Parana, Pinho do Brazil, Pinho Blanio, or Pino.

Parana Pine is a well known timber, but it is not in fact a true pine. A South American timber, it was originally harvested from the state of Parana in Brazil, hence its name. Commercial supplies now come from other countries in South America such as northern Argentina and Paraguay. The tree has also been introduced in plantations to some tropical countries. It grows up to 36m (120ft) high, with bole diameters of up to 1.2m (4ft); most of the bole is free from branches except near the top. (It has an appearance similar to that of the Chile Pine, or Monkey Puzzle Tree, often to be seen growing as an ornament in parks and gardens.) When converted, the sapwood is normally visible and is slightly paler than the heartwood. The heartwood ranges from pale straw brown through to a mid-brown and has often got bright red or brown streaks included in it; these streaks make it easily identified when present. The grain has no clearly defined growth-ring patterns and is usually straight. It has a fine, even texture that is silky to the touch.

This is a mid-weight softwood that averages, when dry, anything from 14 to 18kg (30 to 40lb) per cubic foot. It is not the easiest of softwoods to dry. Distortion is often great and any end splits will inevitably get longer. Extra stickers and weighing down the stacks can be helpful. Not often exported in thick sizes, these are the most difficult to dry. Initial shrinkage after felling is significant; this may contribute to the tendency to split. Parana Pine is rated as similar in strength to European

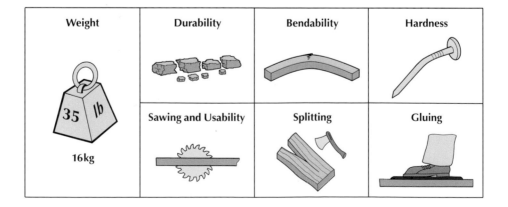

Weight	Durability	Bendability	Hardness
35 lb 16kg			
	Sawing and Usability	Splitting	Gluing

Redwood, but is slightly less flexible. The heartwood is not durable and should be treated with preservatives for external use.

This is a reasonably good material to work. If patchy drying is present in thicker wood there may be some distortion off the saw. It nails reasonably well, but with its tendency to split it is a good idea to pre-drill for all fixings. It glues and stains well and can be polished to a satisfactory finish.

Parana Pine is used for all types of constructional work, joinery, shopfitting and furniture. It is valued at source for its long, clear lengths that are used at home and abroad for staircase manufacture. It is also peeled in Brazil for plywood.

Radial

Tangential

Source:

South America.

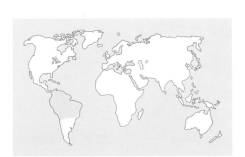

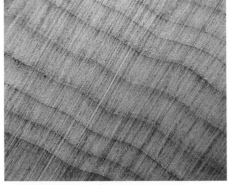

Cross Section

PINE, HARD

Pinus sylvestris

Commonly known as Scots Pine, but also marketed and sold as European Redwood or Redwood, often prefixed with its country of origin. Other names include: Fir, Norway Fir, Scots Fir, simply Red or Yellow and also Red or Yellow Deal.

In addition to Scots Pine, there are a number of other members of the species that can be called 'Hard' Pines. Two of the most common are Radiata Pine (*P. radiata*) and Corsican Pine (*P. nigra*). Both of these, like Scots Pine, have been introduced successfully into many countries worldwide. The overall gross features of each will depend upon local growing conditions, therefore in this description the focus is upon *P. sylvestris* from European sources.

Scots Pine has been introduced into suitable growing climates worldwide and is classed as a hard pine of medium weight. In Europe, the trees will grow to between 30m (100ft) to 43m (140ft) tall if allowed to mature. They will probably have clear boles for two-thirds of this and diameters of 60 to 90cm (2 to 3ft). Exceptional trees have been recorded growing well over 46m (150ft) tall. When grown in plantations they are thinned and cropped at regular thirty to forty year cycles, depending upon local growing conditions and climate. Growth-ring patterns are clearly visible but the amount of darker latewood is much less than that associated with the Pitch Pines. The heartwood has a reddish brown tinge, from which derives the reference to red in its list of alternative names. The sapwood is apparent, showing as a lighter colour. The wood is resinous but not intrusively so, although some

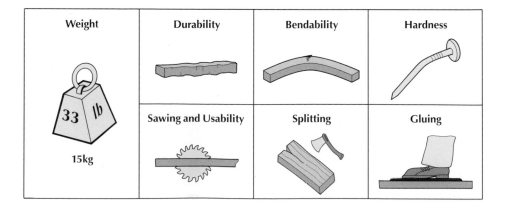

Weight	Durability	Bendability	Hardness
33 lb / **15kg**			
	Sawing and Usability	Splitting	Gluing

resin ducts will often be present. The grain is usually straight and has a medium to coarse texture.

Scots Pine has an average dry weight of between 14 to 16kg (30 to 35lb) per cubic foot. It dries fairly rapidly and well, although some weighting of stacks is advisable to avoid too much distortion. The timber, as European Redwood, is used as a strength comparison for other softwoods. It is classed as non-durable.

All the working qualities associated with Scots Pine are good so long as tools are kept sharp. It glues, stains and paints well. Where knots are present, it is advisable to treat with 'knotting' to avoid any resin seeping through.

This timber is used for most constructional, joinery, moulding and shopfitting applications, there are not many applications like this that it can not be used for.

Source:

UK and Continental Europe.

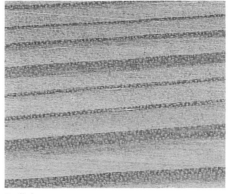

Radial

Tangential

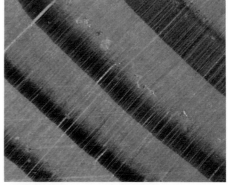

Cross Section

PINE, PITCH

Pinus spp.

The most common in this group are: Longleaf Pine (*P. palustris*) and Slash Pine (*P. elliottii*). Both these are often marketed under the names: American Pitch Pine, Florida Longleaf Pine, Florida Yellow Pine, Georgian Yellow Pine, Gulf Coast Pitch Pine, Longleaf Pitch Pine, Longleaf Yellow Pine, Southern Pine, and Southern Yellow Pine. Included also is Caribbean Pitch Pine (*P. caribaea*). Also known as: Caribbean Longleaf Pine and can be prefixed by a country of origin, such as Guatemala Pitch Pine.

This is a grouping of the heavier, harder Pines that are most often known as 'Pitch Pine'. It is generally accepted that *P. palustris* is probably the best known and has been the most widely used of this group; this wood is not now readily available. All are longleaf Pines with needles up to 18 to 20cm (7 to 8in) long. The trees will grow up to 30m (100ft) tall in good conditions with most of the branches near the top. Clear boles of 15m (50ft) or more are common, with diameters of around 1m (3ft). When cut, the wood is obviously resinous and gives off an odour that remains with it. The timber has very prominent growth rings that clearly distinguish the lighter earlywood from the darker latewood. Sometimes the grain pattern associated with the early and latewood can be curly and this, with the contrasting colours, makes it very attractive. The sapwood tends to be slightly paler and may blend into the heartwood.

These are hard and heavy Pines with average dry weights of anything from 18 to 21kg (40 to 46lb) per cubic foot. They are slow dryers and especially thick material will have a tendency to split and surface check.

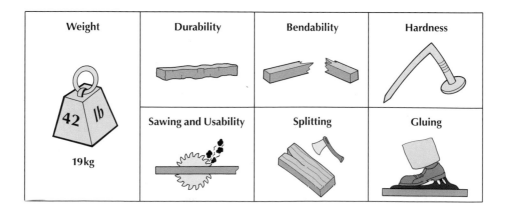

Weight	Durability	Bendability	Hardness
42 lb 19kg			
	Sawing and Usability	Splitting	Gluing

They are extremely strong, stiff and have good shock-resistance properties. Surprisingly, they are considered to be only moderately durable and should be treated for external use, although this may be difficult!

Although the resins can clog up both saws and cutters, the timber will machine reasonably well. Because these are hard woods it is useful to pre-drill for all fixings; they glue reasonably well. Unlikely to be stained, they will need sealing before painting; a satisfactory surface finish can be achieved.

Pitch Pine is used for all types of structural and constructional applications. Traditionally used to make furniture in churches and schools, it also makes an excellent flooring material.

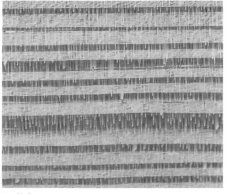

Radial

Tangential

Source: Predominantly North, Central and South America (they have been introduced into other countries).

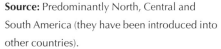

Cross Section

207

PINE, SOFT

Pinus spp.

The most common in this group are:

P. strobus, Yellow Pine, also known as Cork, Pattern, Pumpkin, Quebec, Sapling, Soft, White, Eastern White or Northern White and Weymouth Pine, *P. monticola*, Idaho Pine or Western White Pine or *P. lambertiana*, California Sugar Pine or Sugar Pine.

Within this group fall some of the largest recorded growths for Pines. Sugar Pine is probably the tallest, closely followed by Yellow Pine. Both can attain heights at maturity of well over 46m (150ft) in favourable growing conditions. However, they are relatively soft and light, putting them at the opposite end of the spectrum from the Pitch Pines. When sawn, the sapwood stands out as a slightly lighter colour than the heartwood. Growth ring patterns are apparent although not nearly so distinct as in other Pines; of the three,

Western White Pine has the most pronounced. The heartwood is a light straw or brown colour with the occasional pink tinge. Fine darker lines, or resin ducts, are often seen on the surface of the wood. The grain is usually straight, fine and of a mild texture.

The dry weight of this group averages no more than 11kg (24lb), per cubic foot. Somewhat weaker than European Redwood, these Pines are easily dented, all are brittle and not very shock or split resistant. Western White is probably slightly stronger than Yellow Pine. They will dry rapidly and well with little shrinkage or further movement, but some brown staining can occur if care is not taken. They are rated as a non-durable group of timbers and should be treated with preservatives if used in external applications.

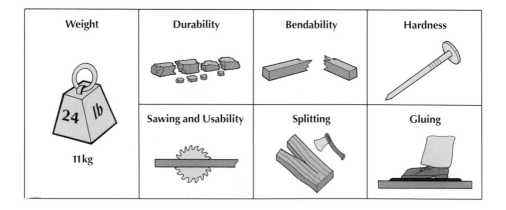

Weight	Durability	Bendability	Hardness
24 lb / 11kg			
	Sawing and Usability	Splitting	Gluing

Soft Pines work well with little blunting of tools. Keep planer knives and cutters sharp to avoid woolly surfaces and the possibility of 'crumbling' at edges. They nail and screw easily, but watch that screws do not pull through! They glue well, take stains evenly and can be painted with ease. Final polished surfaces can be dented if care is not taken.

Top quality soft Pines are sought after for pattern-making because of both their stability and ability to hold a cut edge. They are used for cabinet-making, panelling, window blinds and furniture. Parts of organs and other musical instruments can be produced from these Pines. They are very popular for making a whole host of small mouldings and internal joinery products.

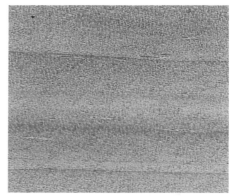

Radial

Tangential

Source:

Predominantly North America.

Cross Section

SPRUCE, EUROPEAN

Picea abies

Also known as: White Deal, Common or Norway Spruce, Whitewood prefixed by Baltic, Finnish, Russian, and so on.

European Spruce is one of the principal timber-producing trees to be marketed, with others as 'Whitewood'. Technically speaking, the use of the word Deal as an alternative name is incorrect. This is an old-fashioned phrase that refers to board size and not to the timber type. It is a native European tree, but is grown in other similar climates as well. Preferring hilly, poor soil locations, in favourable conditions it can grow up to heights of 46m (150ft), with bole diameters of up to 1.2m (4ft). However, most will be cropped from plantations or natural forests well before they attain maximum size, probably on a thirty, or more, year cycle. Small trees are easily recognised as 'Christmas' trees. Timber from European Spruce is nearly white, with little or no difference between the heart and sapwood. Growth-ring patterns are apparent but not darkly defined. Although there may be a lot of resin showing on the bark of the growing trees, little is noticeable in the lumber. There is no heavy, distinctive odour attached to the wood. Knots may be included and will stand out darkly against the pale wood background; the grain is usually straight and of a fairly fine texture.

A mid-weight timber that, averages when dry around 13kg (29lb) per cubic foot. Slightly lighter than European Redwood, it has similar strength values. It will dry fairly rapidly, but probably with some distortion. If economics allow, stacks and kiln charges should be weighted

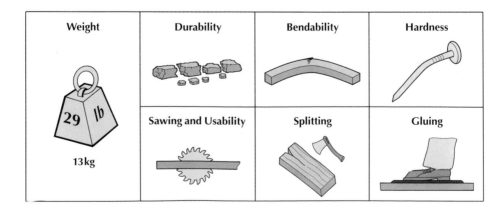

Weight	Durability	Bendability	Hardness
29 lb / 13kg			
	Sawing and Usability	Splitting	Gluing

down to help minimize this. Once dry, it will hold its shape reasonably well. European Spruce is not classified as a durable timber and should be preservative-treated for external use.

This is a good wood to work so long as there are not too many knots or spiral grain. When planed, the surface has a slight sheen with little resin apparent. It nails well and probably only needs pilot holes when screwing. It glues satisfactorily and will stain, albeit patchily at times. It can be painted or varnished, but the knots will need treatment initially.

European Spruce is most often mixed with Silver Fir and sold as Whitewood, then used for internal joinery, mouldings, carcass work, telegraph poles, turpentine and pulp. Top quality, slow mountain-grown material is used in the construction of violins and piano soundboards.

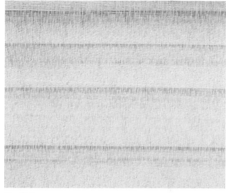

Radial

Tangential

Source:

UK, Continental Europe and similar climates.

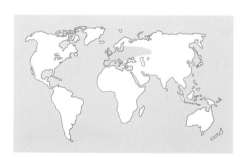

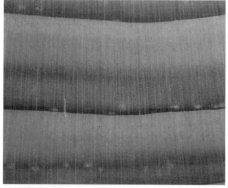

Cross Section

SPRUCE, SITKA

Picea sitchensis

Also known as: Coast, Menzies, Sliver, Tideland, Yellow, West Coast, and Western Spruce.

A native of the west coast of North America, Sitka Spruce has been introduced to many other areas where it can be grown as a plantation crop. In favourable growing conditions trees have been known to top 76m (250ft) high! A more usual size found in forest locations will be around 46m (150ft) with bole diameters of up to 1.5m (5ft). Cropping of plantation-grown Sitka Spruce, outside its country of origin, will probably take place on a forty year, or more, cycle, depending upon the end use of the timber. Thinnings will be taken before this and used for pulp and fencing material. In lumber, the sapwood may only just be discernible, blending gradually into the heartwood. The overall colour of the wood is a pale straw or light yellow colour with some slightly darker growth rings apparent; unlike some of the other Spruces Sitka can have a slight pink hue. If cut on the tangential face it may have a dapple-like appearance. It is a rather bland timber without odour or taste; resins are not apparent. Top quality material will have a straight, even grain with a fine and smooth texture.

Sitka Spruce's dry weight is very similar to its European cousin, but can be lighter when grown in favourable plantation conditions. On average, it will be between 11 to 14kg (24 to 30lb), per cubic foot. It has good strength to weight properties. Although lighter than European Redwood, it is of a similar strength, bending reasonably well and resisting splitting. It is a quick but difficult timber to dry if care is not taken; warping, twisting and loosening of

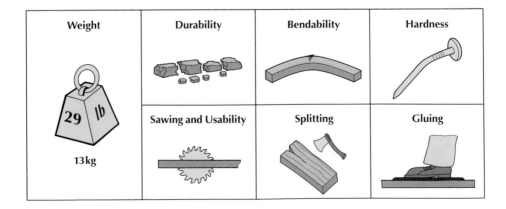

Weight	Durability	Bendability	Hardness
29 lb 13kg			
	Sawing and Usability	Splitting	Gluing

knots are the most common problems. Weigh down stacks and kiln charges if possible. The wood is not considered to be durable and should be treated with preservatives if used for external applications.

The working properties of this wood are all very similar to those of European Spruce. Knots can be troublesome and all tools should be kept sharp.

Forest-grown timber from North America is used for joinery, carcass and cabinet work and internal mouldings. Top quality stuff is used in small aeroplanes, gliders and in musical instruments. Plantation wood, thinnings and so on, are used for packing cases, pallets, fencing and other agricultural uses. Sitka Spruce is often cultivated as a base material for pulpwood in paper production.

Radial

Tangential

Source: Western coastal regions of North America originally and introduced to the UK, Continental Europe and similar climates.

Cross Section

213

YEW

Taxus baccata

Also known as: Common, Irish or European Yew.

Yew is one of the trees that derives from antiquity and it can live for long periods. It is most often seen planted in church yards to ward off evil spirits! Originally thought to be one tree with differing growth patterns, depending upon locality, it is now understood that there may be several varieties. Although it is classified as a softwood, timber from the tree is hard. The Yew tree, in various shapes and sizes, is found extensively throughout Europe; in limited amounts it is also found in Asia and North Africa. The tree is not often grown commercially, taking a long time to reach maturity, but it can reach heights of up to 27m (90ft). Such heights are rare and most trees will only attain a growth of about 12 to 15m (40 to 50ft). Bole sizes can vary, and will inevitably be misshapen, twisted and damaged. The typical growing cycle for Yew means that there are irregular growth-ring patterns showing on the wood; this adds to the attractiveness of the overall appearance. The sapwood is white or cream in colour, fairly narrow and clearly visible. The heartwood starts out as anything from a bright orange-red, purple-red or red-brown colour; indeed, once seen, this wood is never forgotten! After exposure, it will darken to a more uniform, warm shade of brown. The grain tends to be irregular or shallowly interlocked, providing an added feature. It has a tight, fine-grained texture that is pleasing to the touch.

This is a heavy softwood that can vary widely in overall weight. On average, when dry, it should be around 19kg (42lb) per cubic foot. It is hard, tough and resistant to wear.

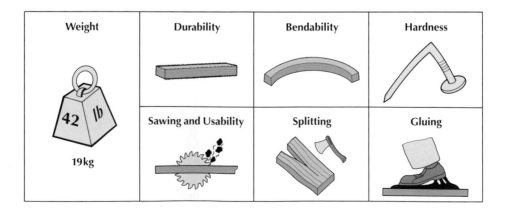

Weight	Durability	Bendability	Hardness
42 *lb* / 19kg			
	Sawing and Usability	Splitting	Gluing

Unbelievably it dries quite quickly and without too much degrade, apart from splits that tend to get bigger. The heartwood is classified as durable.

Yew is a nice wood to work, provided that the grain is not too irregular or interlocked; always keep tools sharp. It will split, so pre-drill when fixing. The slightly oily nature of the wood tends to make gluing difficult at times. It stains well and produces an excellent finished surface.

One traditional use of Yew is in the manufacture of archery bows, especially the English longbow. It turns exceedingly well for fancy goods. Burr Yew is often cut into veneers and is used extensively for reproduction furniture. This is a beautiful wood to end with!

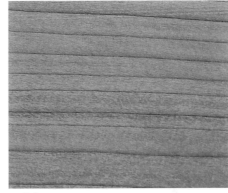

Radial

Tangential

Source:

Throughout Europe, North Africa and into Asia.

Cross Section

215

APPENDIX

WORKING WITH WOOD AND SOME ASSOCIATED HEALTH HAZARDS

In Chapter One, I explained that within the descriptions of each of the individual woods no reference had been made to specific side-effects when working with timber such as skin or eye irritations. In this Appendix I will briefly summarize some of the general conditions that may be caused or aggravated by contact with wood and associated waste from it. It is recognized by the Health and Safety Executive (HSE), in the UK that 'all wood dust is hazardous to health'. Please note that no differentiation is being made by me, or the HSE, between hardwoods or softwoods. Rare forms of nasal cancer, skin irritations and damage to the respiratory system can result if the appropriate precautions are not observed. Each of us may be affected to a lesser or greater degree; the objective of these notes is to raise awareness to ensure that preventative action is taken before rather than after the event.

Those of us who have worked with this wonderful material will be aware of occasions when a splinter has turned septic. This may have been caused by a simple secondary infection from bacteria entering the wound, from some treatment applied to the wood or through the natural toxin present within the species being worked. When a splinter has punctured the skin common sense prevails; remove it as quickly as possible and clean up the wound.

Skin irritations may be caused by wood dust, bark, sap or other forms of low life that may have grown up on it. If a user has a history of dermatitis, appropriate precautions should be made prior to contact. In most cases a form of nettle rash or itching will provide the clue that something is reacting negatively. Sometimes these symptoms will take several days to develop and may persist for some time, or at least for as long as the exposure continues. Covering exposed areas and using efficient dust extraction will alleviate many problems, but failing that change the species and use something else!

Dust probably has the most detrimental effect on our health and can lead, in some rare cases, to nasal cancer. Other more common effects will be blocked and runny noses, violent sneezing and occasional nose bleeding. As with the skin irritations, these latter minor side-effects will generally diminish and disappear in time after use of the particular wood has ceased. If a woodworker has asthma, particular care needs to be taken to ensure that effective controls of dust levels are maintained. Some wood dust can cause a specific allergic reaction. When this occurs, the individual may react even when exposed to limited amounts of similar dust. Eyes often show symptoms of irritation when exposed to dust. The most common will be soreness, watering and, in some cases, conjunctivitis.

One of the major ways to reduce the risk of side-effects when working with wood is to ensure that an efficient dust and chipping extraction system is employed. Those involved in the UK wood-processing industry will be well aware of the regulations they have to comply with under the Control of Substances Hazardous to Health (COSHH). In these, Maximum Exposure Limits (MELs), are set out to control and define how much dust employees can be safely exposed to. Generally, most larger companies are aware of their responsibilities and tend to comply with the requirements. In smaller concerns and with a lot of DIYers this is not often the case. If suitable extraction equipment proves too expensive, then serious consideration should be given to the use of appropriate Respiratory Protective Equipment (RPE). This type of equipment is not as effective as extraction at source but will certainly reduce risk if nothing else is available. With dust, it is what you cannot see that is most likely to have ill effects, so please take every precaution possible to maintain your health.

The effects of toxic wood elements on individuals, businesses and their employees should be taken seriously. In the UK, the HSE produces a host of Information Sheets covering the subject. The list that follows is not necessarily fully comprehensive, and for up-to-date information contact the HSE directly at the addresses provided.

HSE INFORMATION SHEETS

Woodworking sheet number:

1 Wood Dust: Hazards and Precautions.

6 COSHH and the Woodworking Industries.

11 Hardwood Dust Survey.

12 Assessment and Control of Wood Dust: Use of the Dust Lamp.

14 Selection of Respiratory Protective Equipment Suitable for Use with Wood Dust.

30 Toxic Woods.

Further information

- HSE priced and free publications are available by mail order from HSE Books, PO Box 1999, Sudbury, Suffolk CO10 6FS. Telephone: 01787 881165, Fax: 01787 313995.

- HSE priced publications are also available from booksellers.

- British Standards are available from BSI Sales and Customer Services, 389 Chiswick High Road, Chiswick, London W4 4AL. Telephone: 0181 996 7000; Fax: 0181 996 7001.

- For other enquiries ring the HSE's InfoLine, Telephone: 0541 545500, or write to the HSE's Information Centre, Broad Lane, Sheffield S3 7HQ.

- HSE home page on the World Wide Web: http://www.open.gov.uk/hse/hse home.htm.

- OFI have produced a CD-ROM database of wood properties called "Prospect" that is designed to provide information on timber species from all areas of the world. Developed over the last 15 years the system records data from over 1550 species, 92 timber properties and 175 end-uses from over 1800 literature-based references. For more information contact the OFI at the University of Oxford.

INDEX

Entries in **bold** type denote the 100 woods featured in the book. Entries in *italic* type denote all Latin name references. All other entries are incidental mentions of particular woods and alternative names referred to throughout the text. The index is divided into separate sections for hardwoods and softwoods.